Topics in
Current Physics

24

Topics in Current Physics   Founded by Helmut K. V. Lotsch

Amorphous Solids

Low-Temperature Properties

Edited by W. A. Phillips

With Contributions by

A. C. Anderson   B. Golding   J. E. Graebner
S. Hunklinger   J. Jäckle   W. A. Phillips
R. O. Pohl   M. v. Schickfus   D. L. Weaire

With 72 Figures

Springer-Verlag Berlin Heidelberg New York 1981

W. Andrew Phillips, Ph. D.

Cavendish Laboratory, University of Cambridge,
Maddingley Road, Cambridge CB3 OHE, United Kingdom

ISBN-13:978-3-642-81536-2   e-ISBN-13:978-3-642-81534-8
DOI: 10.1007/978-3-642-81534-8

Library of Congress Cataloging in Publication Data. Main entry under title: Amorphous solids. (Topics in current physics ; 24) Bibliography: p. Includes index. 1. Glass—Thermal properties. 2. Amorphous semiconductors—Thermal properties. 3. Materials at low temperatures. 4. Tunneling (Physics) I. Phillips, William Andrew, 1944– II. Series. TA450.A52   536′.56   80-29294

2153/3130-543210

Preface

It is now ten years since it was first convincingly shown that below 1 K the thermal conductivity and the heat capacity of amorphous solids behave in a way which is strikingly different to that of crystalline solids. Since that time there has been a wide variety of experimental and theoretical studies which have not only defined and clarified the low temperature problem more closely, but have also linked these differences between amorphous and crystalline solids to those suggested by older acoustic and thermal experiments (extending up to 100 K). The interest in this somewhat restricted branch of physics lies to a considerable extent in the fact that the differences were so unexpected. It might be thought that as the temperature, probing frequency, or more generally the energy decreases, a continuum description in which structural differences between glass and crystal are concealed should become more accurate. In a sense this is true, but it appears that there exists in an amorphous solid a large density of additional excitations which have no counterpart in normal crystals. This book presents a survey of the wide range of experimental investigations of these low energy excitations, together with a review of the various theoretical models put forward to explain their existence and nature.

There is a danger that with a book of this kind, which attempts to present a coherent survey of an area of active research, articles may be overtaken by the course of events. Even though the chapters have taken longer to assemble than expected this has not proved to be the case. The last two years or so, during which time the manuscripts were completed, have not seen any major developments. In my view this period has been a natural pause or transition between establishing the basic experimental facts and phenomenological theories, and the development (which lies in the future) of clear microscopic descriptions of the excitations. It has therefore been a fortuitously appropriate time to produce this book.

Following two introductory chapters which serve to introduce basic ideas and to put the problem into perspective, three chapters describe measurements of the heat capacity, thermal expansion and thermal conductivity. These measurements define the basic properties of the excitations. More specific acoustic, dielectric and optical measurements are described in Chaps.6, 7 and 8. The general organization of the book is such that the basic physical ideas are introduced in the earlier chapters and followed by more complete discussions as needed later in the book.

Finally, I would like to thank Dr. H. Lotsch of Springer-Verlag (and one or two of the contributors) for the patience shown in awaiting the book, and also the members of the Centre de Recherches sur les Très Basses Températures for their hospitality during the time that the manuscript was completed.

Cambridge, January 1981   *W. Andrew Phillips*

Contents

List of Contributors

Anderson, Ansel C.
> Department of Physics, University of Illinois,
> Urbana, ILL 61801, USA

Golding, Brage
> Bell Laboratories, 600 Mountain Avenue,
> Murray Hill, NJ 07979, USA

Graebner, John E.
> Bell Laboratories, 600 Mountain Avenue,
> Murray Hill, NJ 07974, USA

Hunklinger, Siegfried
> Max-Planck-Institut für Festkörperforschung, Büsnauer Straße 171
> D-7000 Stuttgart 80, Fed. Rep. of Germany

Jäckle, Josef
> Fakultät für Physik, Universität Konstanz
> D-7750 Konstanz, Fed. Rep. of Germany

Phillips, W. Andrew
> Cavendish Laboratory, University of Cambridge
> Madlingley Road, Cambridge CB3 OHE, United Kingdom

Pohl, Robert O.
> Laboratory of Atomic and Solid State Physics, Clark Hall,
> Cornell University, Ithaca, NY 14853, USA

von Schickfus, Manfred
> Max-Planck-Institut für Festkörperforschung, Büsnauer Straße 171
> D-7000 Stuttgart 80, Fed. Rep. of Germany

Weaire, Denis L.
> Physics Department, University College, Belfield,
> Dublin, Ireland

1. Introduction

W. A. Phillips

With 1 Figure

The choice of a title for this book presented considerable difficulty. The subject matter, on the other hand, did not, and can be summarized as follows. Measurements of thermal, acoustic and dielectric properties of amorphous or disordered solids below 50 K show a number of unusual features which have no counterpart in corresponding measurements in crystals. Similarly, optical measurements at characteristic frequencies below 10^{12}Hz show marked differences between amorphous and crystalline solids. It is these differences that provide the subject matter.

The experiments to be described here have been performed at temperatures varying from 10 mK to room temperature and so the "low temperature" in the title is not strictly appropriate. There is clearly a characteristic energy of less than 5 meV associated with the experiments but the use of "low energy" would not indicate that electronic excitations were excluded. Similarly "low frequency" is misleading, as it is the essence of many of these effects that the excitations cannot be characterised by a well-defined frequency. The title chosen is a compromise and agrees with common usage.

The first introductory chapter begins in Sect.1.1 with a short historical account followed in Sect.1.2 by a discussion of the quantum mechanics of a particle in a double-well potential, for reasons made clear in Sect.1.1. Finally, Sect.1.3 outlines the subject matter and approach of the remainder of the book.

1.1 Historical Background

It is roughly ten years since ZELLER and POHL [1.1] presented clear and unambiguous evidence that below 1 K the thermal properties of amorphous insulating solids differ remarkably from their crystalline counterparts. Earlier measurements, some of which are mentioned in Chap.5, were not taken as seriously as they should have been, but the work of ZELLER and POHL triggered off a wide range of experimental studies and theoretical suggestions. This immediate interest in the subject is easy to understand, as the thermal properties of crystalline insulators in the temperature range below 1 K present no surprises and are easily understood in terms of the Debye theory. Before 1970 most solid-state physicists, if asked to describe the behaviour of the heat capacity and the thermal conductivity of pure fused silica below 1 K, would have predicted a behaviour similar to that of crystalline quartz on the grounds that the structural irregularities in glasses become progressively less important as the phonon wavelength increases.

A second attractive feature of the results was the apparent universality of behaviour. The initial measurements showed that not only oxide glasses but also inorganic (selenium) and organic polymers had a heat capacity which varied approximately linearly in temperature T and a thermal conductivity which varied as T^2. Further, the absolute values of the two parameters lay within an order of magnitude for all amorphous solids. This feature, in particular, was an immediate magnet for theorists.

Within three years about ten different models had been proposed to account for these effects. In roughly chronological order the basic ideas behind these theories can be listed as follows: localised electron states [1.2], dispersive or damped phonons [1.3,4], tunneling states [1.5,6], cavity models [1.7], cellular and microcrystallite effects [1.8,9], scattering from structural inhomogeneities [1.10,11], dislocations [1.12,13], and "propagating bosons" [1.14]. About half these theories provided explanations for both the thermal conductivity and the heat capacity, the remainder concentrating on one or the other. Although there were problems in reconciling the cellular and microcrystallite models in particular with structural studies, in general thermal studies provided an insufficient basis for differentiation between theories.

The publication of the results of acoustic studies beginning in 1973 [1.15,16] changed the picture dramatically. As described in Chaps.6 and 7 these experiments showed that the magnitudes of the thermal conductivity in oxide glasses could be deduced from the scattering of acoustic phonons, and also that the acoustic attenuation could be saturated. This provides strong evidence that Debye-like phonons exist in glasses, and are scattered by additional excitations which can be represented most simply by two-level systems, or more generally by highly anharmonic oscillators. For this reason most experimental work has since that time been interpretated in terms of the two-level system model, which in turn can be related to the tunneling model. None of the other models can adequately explain both the thermal and the acoustic data, although of course there is no a priori reason why all the phenomena should be closely linked.

The tunneling model, whether correct or not, has provided a basic theoretical framework for the interpretation of experimental studies: experiments are claimed either to support or to disprove this model. The basic idea behind the model is that in a disordered solid, as opposed to a crystalline one, certain atoms or groups of atoms have available two (or possibly more) mutually accessible potential minima. These atoms therefore move in a potential of the form shown in Fig.1.1. The position co-ordinate is often referred to as a configurational co-ordinate to

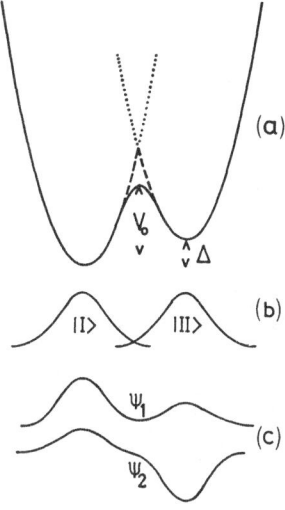

Fig.1.1. (a) A double-well potential built up from two harmonic-oscillator potentials (continued as dotted lines) either cut off sharply at the point of intersection (dashed line) or joined smoothly (solid line). (b) The two states |I> and |II> used in the localized representation. (c) The lowest two eigenstates ψ_1 and ψ_2 of the double-well potential

conceal the fact that the microscope nature of the state is unknown. At low temperatures only the two lowest energy levels are important, with an energy difference determined by quantum mechanical tunneling through the barrier, and by the asymmetry Δ. The wide range of local potentials present in the disordered solid gives rise to a wide range of possible energy differences, which can be represented by an approximately constant density of states. This model is an extension of that used to explain the properties of certain impurities in alkali halide crystals [1.17] but has also been justified as a representation of the excess entropy found in glasses when compared to the corresponding crystal [1.18]. A quantum mechanical treatment of the double well is presented in Sect.1.2.

Returning to the historical development, two key experiments showed that the formal analogy between a two-level system and a spin 1/2 particle in a magnetic field was not an empty one. The observation [1.19] of a linewidth larger than that calculated from the direct interaction between phonons and two-level systems showed the need for two relaxation times, denoted T_1 and T_2 by analogy with the magnetic case, to describe the properties of the two-level system, and also suggested that interactions occurred between systems. The observation [1.20] of phonon echoes showed the importance of coherent effects, and also provided a precise way of measuring T_1 and T_2. Subsequent experiments and theory, described in Chap.7, have given detailed information on the strength of the coupling between phonons and two-level systems, but in general these phonon experiments have not given any clear indication of the microscopic nature of the systems.

Concurrently with the acoustic work a wide range of different materials was being investigated by more traditional experiments. The range of materials showing the thermal anomalies below 1 K was extended to include disordered crystals [1.21] and metallic glasses [1.22], thus showing not only the generality of the effect,

but also that localized disorder, as present in imperfect crystals, was sufficient
to produce the effects found in true amorphous or glassy solids. Further, both
thermal [1.23] and dielectric [1.24] measurements showed that impurities could influence the properties of glasses below 1 K.

Our present knowledge of these low temperature effects in glasses is summed up
in the remainder of this book. It is probably fair to say that although a great
deal is known about the macroscopic behaviour of amorphous and disordered solids
at low temperatures, and a good phenomenological description of this behaviour is
available, much less is known about the microscope basis for this behaviour. The
acoustic (and equivalent electric) experiments are well described by the two-level
system model, but there is no general agreement that this is the same as the tunneling state model, which in itself does not provide a specific microscopic description applicable to a particular amorphous solid. This detailed microscopic
picture is the most important remaining problem.

1.2 Tunneling States

1.2.1 Energy Levels

The quantum mechanical problem of a particle moving in a double-well potential is
not a difficult one. It is, however, difficult to find a reasonably straightforward but complete account, as the problem appears to fall between the class of
those that are automatically included in textbooks, and those that are discussed
in detail in research articles. For this reason the solution is presented here as
an introduction to many of the results quoted later in the book. An attempt has
been made to draw attention to one or two rarely discussed problems.

The calculation of the energy levels of a particle in a double-well potential
V of the form shown in Fig.1.1a usually starts with the solution of the single-well
problem shown in Fig.1.1b. The choice of these two basis states is known as the
well, non-diagonal or *localized* representation. Each state is the ground state of
the appropriate harmonic potential V_1 or V_2, both of which are shown continued as
dotted lines in Fig.1.1a. The Hamiltonian can be written as

$$H = H_1 + (V - V_1) = H_2 + (V - V_2) \tag{1.1}$$

where H_1 and H_2 are the individual Hamiltonian operators. In this representation
the Hamiltonian matrix becomes

$$\begin{pmatrix} E_1 + <1|V - V_1|1> & <1|H|2> \\ <2|H|1> & E_2 + <2|V - V_2|2> \end{pmatrix} \tag{1.2}$$

To a good approximation each term $<i|V - V_i|i>$ can be neglected in comparison with E_i. If the zero of energy is chosen as the mean of the two ground state energies E_1 and E_2, (1.2) can be written

$$\frac{1}{2}\begin{pmatrix} -\Delta & -\Delta_0 \\ -\Delta_0 & \Delta \end{pmatrix} \tag{1.3}$$

where Δ_0 is defined as

$$\Delta_0 = -2<1|H|2> , \tag{1.4}$$

the negative sign being introduced because the matrix element $<1|H|2>$ is negative. Notice that only if the wells are identical, apart from a relative displacement in energy, is the Δ of (1.3) identical to that of Fig.1.1a.

Δ_0 can be evaluated for specific potentials. For two identical three-dimensional harmonic oscillators with $\Delta = 0$ and with an overall potential shown by the dashed continuation in Fig.1.1a

$$\Delta_0 = \hbar\omega_0\left[3 - \left(\frac{8V_0}{\pi\hbar\omega_0}\right)^{\frac{1}{2}}\right]e^{-2V_0/\hbar\omega_0} \tag{1.5}$$

where V_0 is the minimum energy barrier between the two wells, and $\hbar\omega_0$ is equal to $2E_1$ or $2E_2$. Since for our purposes (and indeed for the localised representation to be useful) $V_0 \gg \hbar\omega_0$, (1.5) becomes

$$\begin{aligned}
\Delta_0 &= -2\hbar\omega_0\left(\frac{2V_0}{\pi\hbar\omega_0}\right)^{\frac{1}{2}}e^{-2V_0/\hbar\omega_0} \\
&= -4\left(\frac{2V_0^3\hbar^2}{md^2}\right)^{\frac{1}{4}}e^{-(2mV_0/\hbar^2)^{\frac{1}{2}}d/2}
\end{aligned} \tag{1.6}$$

where m is the mass of the particle and d the separation of the two minima. This value of Δ_0 is just twice that found for the equivalent problem in one dimension [1.25].

As an alternative example, the solution of Mathieu's equation for a rigid rotator in a twofold symmetric potential gives [1.26], in the same limit $V_0 \gg \hbar\omega_0$, the approximate result

$$\Delta_0 \cong -8\hbar\omega_0\left(\frac{2V_0}{\pi\hbar\omega_0}\right)^{\frac{1}{2}}e^{-4V_0/\hbar\omega_0} \tag{1.7}$$

where again V_0 is the height of the barrier separating the two wells, and ω_0 is the angular frequency of small oscillations within a single minimum.

Although both results imply a similar exponential dependence of Δ_0 on V_0 the numerical relationships are different, and the use of a particular form of equation for Δ_0 cannot be justified unless, as is often the case in crystals, the microscopic symmetry of the potential is known. For this reason an expression of the form

$$\Delta_0 = -\hbar\Omega e^{-d(2mV_0/\hbar^2)^{\frac{1}{2}}}$$

(1.8)

where $\hbar\Omega$ is an energy roughly equal to $\hbar\omega_0$, is usually adequate in the case of amorphous solids. It is worth noting that $<1|H|2>$ is negative in these examples because the negative contribution from $V-V_2$ (for example) in (1.1) overwhelms the positive contribution from H_2 (in contrast to the suggestion of 1.27).

The matrix (1.3) can be diagonalized to obtain the eigenstates, the *true, diagonal* or *energy* representation. The eigenfunctions, shown in Fig.1.1c, have energies $\pm E/2$ where

$$E^2 = (\Delta^2 + \Delta_0^2)$$

(1.9)

and are usually written in the analytic forms

$$\psi_1 = |1>\cos\theta + |2>\sin\theta$$

(1.10)

$$\psi_2 = |1>\sin\theta - |2>\cos\theta$$

(1.11)

where $\tan 2\theta = \Delta_0/\Delta$. ψ_2 is the lower energy state because Δ_0 (as defined here) is positive.

The functions ψ_1 and ψ_2 defined by (1.10) and (1.11) are orthonormal only to the extent that the overlap term $<1|2>$ can be put equal to zero. As far as normalization is concerned this presents no difficulty, as an additional multiplicative factor can be readily calculated, but may lead to more important problems in the calculation of matrix elements. Any matrix element involving the overlap of $<1|$ and $|2>$ must be evaluated carefully through a correct choice of orthonormal states as discussed by SUSSMAN [1.28].

It is worth pointing out the existence of a wide range of notation for (1.9). Not only have several symbols been used for the various energies in this equation but the same symbols have been used for half the energies. Occasional discrepancies of powers of two in formulae cited in the literature can often be traced to this difference in definition. This book uses the notation of (1.9) throughout.

For this, as for any problem involving two energy levels, there is a formal analogy with the problem of a spin 1/2 particle in a magnetic field. The Hamiltonian matrix (1.3) can be rewritten in terms of spin operators, although here too there is a choice of notation between the Pauli spin matrices

$$\sigma_x = \begin{pmatrix} 0 & 1 \\ 1 & 0 \end{pmatrix} \ \sigma_y = \begin{pmatrix} 0 & -i \\ i & 0 \end{pmatrix} \ \sigma_z = \begin{pmatrix} 1 & 0 \\ 0 & -1 \end{pmatrix}$$

(1.12)

and the spin -1/2 operators defined by $S_i = \frac{1}{2}\sigma_i$. After diagonalization the Hamiltonian can be written in the obvious form

$$H = \frac{1}{2} E\sigma_z \ .$$

(1.13)

The advantages of this analogy are seen most clearly in Chaps.6 and 7 where it is used to interpret non-linear and coherent effects in the interaction of the two levels with acoustic and electric fields. A detailed discussion is given at the appropriate point; in the meantime it is sufficient to give a treatment based on the Einstein A and B coefficients to calculate the expressions for relaxation times and phonon scattering rates needed in earlier chapters.

1.2.2 Transition Probabilities and Relaxation Times

Transitions between the states ψ_1 and ψ_2 occur through the perturbation of the potential well of Fig.1.1 a by a photon or phonon with energy $\hbar\omega = E$. This perturbation can change Δ or Δ_0 (or both) but, of the two, changes in Δ are much more important.

The reason for this is twofold. The first is that the wavelength of the perturbing electric or strain field is much greater than the separation of the wells. As in the electric dipole approximation in semiclassical radiation theory, this leads to a perturbing potential which is essentially antisymmetric, equivalent to a change in Δ and not in Δ_0. Secondly, the matrix elements, calculated in the localised basis $|1>$ and $|2>$, are relatively much smaller for a symmetric perturbation. (The use of the words symmetric and antisymmetric is of course not exactly correct, because the potential well of Fig.1.1a is not symmetric. It is, however, a useful approximation, identifying perturbations that tend to change Δ or Δ_0 separately).

This second effect can be illustrated by comparing the matrix elements for two perturbing potentials Ax and Bx^2 for the one-dimensional double harmonic oscillator of Fig.1.1a with $\Delta = 0$. The appropriate quantity for comparison is the ratio of the matrix element to the change in energy of each well (Ad/2 and $Bd^2/4$ in the two cases, where d is the separation of the minima). For the antisymmetric perturbation this ratio is ±1 for the diagonal matrix elements and zero for the off-diagonal. In the case of the symmetric perturbation Bx^2 the diagonal elements are equal in both magnitude and sign and so give simply a shift in the zero of energy, while the ratio of the off-diagonal matrix elements to $Bd^2/4$ is $(\hbar\omega_0/4V)\exp(-2V_0/\hbar\omega_0)$. $V_0/\hbar\omega_0$ is considerably greater than unity for the low temperature applications of this model, and so the off-diagonal terms in the $|1>$, $|2>$ basis can be neglected [notice that for $\Delta = 0$ the functions ψ_1 and ψ_2 defined by (1.10) and (1.11) are orthogonal, and so no particular precautions need be taken in the calculation].

This result has general validity. If the potential wells are not equivalent, or if Δ is not equal to zero, the antisymmetric perturbation will give off-diagonal terms and the symmetric perturbation will give, in addition to the off-diagonal terms, unequal diagonal terms. However, all these matrix elements are proportional to a factor of the general form $\exp(-2V_0/\hbar\omega_0)$ and so will be relatively unimportant. The perturbation to be included in the Hamiltonian (1.3) is therefore diagonal in the basis $|1>$, $|2>$. Using the transformation defined by (1.10) and (1.11), the perturbation in the ψ_1, ψ_2 basis has the form

$$\begin{pmatrix} \cos 2\theta & \sin 2\theta \\ \sin 2\theta & -\cos 2\theta \end{pmatrix} \quad \text{or} \quad \begin{pmatrix} \Delta/E & \Delta_0/E \\ \Delta_0/E & -\Delta/E \end{pmatrix}.$$

The interaction Hamiltonian can therefore be written in terms of the Pauli operators as

$$H_{int} = \left(\frac{\Delta}{E} \sigma_z + \frac{\Delta_0}{E} \sigma_x \right) p_0 \cdot F_m + \left(\frac{\Delta}{E} \sigma_z + \frac{\Delta_0}{E} \sigma_x \right) \gamma e \tag{1.14}$$

in the presence of an electric field F and a strain field e. p_0 and γ are defined as $1/2 \partial \Delta / \partial F$ and $1/2 \partial \Delta / \partial e$ respectively, and are therefore also equal to the electric and elastic dipole moments of the equivalent classical potential. The vector character of F is here preserved, but the quantity γe has been written as an average over orientations (it can also be averaged over polarizations, although transverse and longitudinal models are usually considered separately). The off-diagonal term σ_x produces transitions between ψ_1 and ψ_2 while the diagonal term σ_z changes their relative energies.

In many cases the results of experiments can be interpreted in terms of a simpler model which ignores the relationship between the diagonal and off-diagonal terms in (1.14). This *two-level-system* model, so called to distinguish it from the *tunneling* model, uses a single parameter E for the energy difference between the two levels. The interaction Hamiltonian is written in the form

$$H_{int} = \left(\frac{1}{2} D\sigma_z + M\sigma_x \right) e + \left(\frac{1}{2} \mu \sigma_z + \mu' \sigma_x \right) F \tag{1.15}$$

where the diagonal and off-diagonal terms are specified independently [the factors of 1/2 in (1.15) are another source of possible confusion, as is the factor of 1/2 in the definition of p_0 or γ].

The interaction between tunneling states of two-level systems and phonons or photons can conveniently be described by the use of the Einstein coefficients (this treatment resembles that of THOMAS [1.29]). Consider first the interaction of two-level systems with thermal phonons. Each two-level system is continually absorbing and emitting thermal phonons. The rate equation for the probability p_1 of finding the system in the ground state ψ_1 can be written

$$\frac{dp_1}{dt} = -p_1 B\rho(E) + p_2[A + B\rho(E)] \tag{1.15}$$

where A and B are phonon Einstein coefficients and $\rho(E)$ is the phonon energy density (per unit volume) evaluated at an energy E equal to the two-level system energy. $\rho(E)$ is given in terms of the density of states $g(E)$ by

$$\rho(E) = \frac{Eg(E)}{e^{E/kT} - 1}. \tag{1.16}$$

In thermal equilibrium $dp_1/dt = 0$, and since $p_1 + p_2 = 1$

$$\frac{A}{B} = Eg(E) \quad . \tag{1.17}$$

For small departures from equilibrium (1.15) defines a relaxation time τ, where

$$\tau^{-1} = [A + 2B\rho(E)] \quad , \tag{1.18}$$

which can be rewritten using (1.16) as

$$\tau^{-1} = A \coth(E/2kT) \quad . \tag{1.19}$$

$1/A$ is the natural lifetime of the system at absolute zero, and at any temperature $\hbar\tau^{-1}$ is the uncertainty in the energy E.

The analysis can be continued to calculate the rate at which phonons are scattered by the two-level systems. If the density of states of the two-level systems is $n(E)$ per unit volume, the change in the phonon energy density is given by

$$\frac{\partial\rho(E)}{\partial t} = E\, n(E)\, \frac{\partial p_1}{\partial t} \tag{1.20}$$

so that using (1.15)

$$\frac{\rho(E)}{dt} + n(E)BE(p_1 - p_2)\rho(E) = n(E)EAp_2 \quad .$$

The phonon lifetime is given by

$$\tau_{ph}^{-1} = n(E)BE(p_1 - p_2) = \frac{n(E)A}{g(E)}\, \tanh(E/2kT) \quad . \tag{1.21}$$

The coefficient B can be calculated starting from (1.14) using a derivation equivalent to that of the corresponding optical problem. However, unlike the optical case, the form chosen for the density of states $g(E)$ must be specified for phonons, as must the polarization. As shown in Chaps.3 and 5 the Debye approximation can be used at temperatures of 1 K and below, so that the phonon density of states has a quadratic dependence on energy. Further, the phonon polarization can be classified as either longitudinal l or transverse t. For a single polarization α, B is given by

$$B = \frac{\pi M_\alpha^2}{\hbar\rho_0 v_\alpha^2} \tag{1.22}$$

where ρ_0 is the density of the solid and v_α is the velocity of sound for polarization α. M, as discussed in connection with (1.14), is an average over orientations. The relaxation times can now be written for the *two-level system* as [1.30]

$$\tau^{-1} = \sum_\alpha \left(\frac{M_\alpha^2}{v_\alpha^5}\right) \frac{E^3}{2\pi\rho_0\hbar^4}\, \coth\left(\frac{E}{2kT}\right) \tag{1.23}$$

and

$$\tau_{ph}^{-1} = \frac{\pi M_\alpha^2}{h\rho_0 v_\alpha^2} \, n(E)E \, \tanh\left(\frac{E}{2kT}\right) \tag{1.24}$$

where both longitudinal and transverse contributions have been included in the expression for τ^{-1}, but τ_{ph}^{-1} is written for a single polarization α.

For the *tunneling states* these formulae must be modified to take account of the explicit form of M_α. The tunneling state relaxation time is

$$\tau^{-1} = \sum_\alpha \left(\frac{\gamma_\alpha^2}{v_\alpha^5}\right) \frac{\Delta_0^2 E}{2\pi\rho_0 \hbar^4} \, \coth\left(\frac{E}{2kT}\right) \tag{1.25}$$

but the expression for the phonon scattering time is more complicated because, for a given phonon energy E, each tunneling state scatters phonons at a rate determined by M_α^2, proportional to Δ_0^2. In an amorphous solid, as opposed to a crystal, there will in general be a wide range of local environments and hence a range of values of Δ_0. The exact expression for τ_{ph}^{-1} involves the distribution funtion for Δ_0. This point will be discussed in more detail later in the book and for the moment it is sufficient to write

$$\tau_{ph}^{-1} = \frac{\pi \gamma_\alpha^2}{\hbar\rho_0 v_\alpha^2} \, \bar{n}(E)E \, \tanh(E/k2T) \tag{1.26}$$

where $\bar{n}(E)$ is an effective density of states. In principle, the coupling parameter γ_α also varies from tunneling state to tunneling state, and it too should be represented by an average, although this can be incorporated into $\bar{n}(E)$.

One important difference between the two-level-system model and the tunneling-state model should be noticed. As described in Chap.3 the specific heat can be calculated from $n(E)$. In the two-level-system model the same parameter enters directly into the expression for τ_{ph}^{-1}, but in the tunneling model this is not so, and the relationship between the heat capacity and τ_{ph}^{-1} depends on an unknown distribution function.

A description of absorption and emission in terms of the Einstein coefficients and semiclassical radiation theory is obviously oversimplified, and it is important to consider the extent to which it is valid. The semiclassical approach can be replaced by a quantized field calculation without changing the results. More important is the neglect in the Einstein treatment of coherence between the wave functions of the two energy levels. The two-level system is here characterised by two parameters, the occupation probabilities, instead of by the three that a full quantum treatment requires. Thie limitation means that the Einstein approach cannot provide a detailed explanation of non-linear and coherent effects (including higher-order transitions involved in Raman scattering). It does, however, give an accurate description of one-phonon or photon emission and absorption, although even simple scattering needs to be treated more carefully. In principle, scattering of phonons,

for example, cannot be considered as two independent processes, absorption followed
by emission. This is illustrated by the phenomenon of resonance fluorescence
[1.31] where the width of a scattered beam is the same as that of the incident
beam, and may be less than the natural linewidth of the two-level system. However,
in the case where the incident beam has quasi-continuous spectrum, scattering *can*
be considered as absorption followed by emission. This is the case considered here
in the calculation of τ_{ph}, and so the use of the Einstein coefficients is valid.

1.3 Organization of the Book

This chapter, in addition to providing a brief historical introduction, has presented a fairly detailed discussion of the quantum mechanics of a particle in a
double well as a preliminary to later chapters. Chapter 2 is also introductory,
giving an outline of the methods used to calculate the vibrational density of
states and at the same time providing a summary of the main features of this density of states. This serves not only to introduce ideas and nomenclature used
later in the book, and to put into perspective the frequency range of the modes
studied here, but also as a reminder that the Debye theory, used almost exclusively by those working at low temperatures, has limited applicability.

Chapter 3 provides a survey of specific heat data in a wide range of amorphous
solids, as well as in other disordered solids. This is followed by a critical
examination of the most promising theoretical suggestions. Ideally, this is a pattern that could usefully have been followed throughout the book, but in practice
it is only possible in Chaps.3 and 5 where a variety of theoretical models is
available. In other chapters experimental results have been critically compared
with the only available model.

Chapter 4 continues the description of the homogeneous equilibrium properties
with a survey of the thermal expansion data, although it must be admitted that
these are very limited. However, in view of the bizarre behaviour of the thermal
expansion at very low temperatures it was felt that some discussion should be included. In order to achieve a balanced chapter a certain amount of editorial li-
cence has been necessary to extend the discussion to include higher temperature
effects.

A survey of thermal conductivity in glasses is presented in Chap.5, following
a similar pattern to that of Chap.3. In addition to discussing the temperature
range below 1 K with which the specific heat chapter was concerned, Chap.5 also
discusses the behaviour between 1 and 50 K, where similar theoretical models may
be applicable.

Chapters 6 and 7 describe the results of a wide range of acoustic and dielectric
experiments. The division of this aspect of the subject into two chapters reflects

the large number of such experiments, and the distinction between the two lies in the role played by coherent effects. Chapter 6 concentrates on relaxation effects, observed in the range 1 to 50 K, and the more straightforward resonant interaction between two-level systems and phonons or photons which is studied below 1 K. Chapter 7 describes in detail the phonon and photon-echo experiments, which can be studied only at very low temperatures, below 50 mK.

The final Chap.8 describes and interprets low frequency Raman scattering in glasses.

References

1.1 R.C. Zeller, R.O. Pohl: Phys. Rev. B4, 2029 (1971)
1.2 D. Redfield: Phys. Rev. Lett. 27, 730 (1971)
1.3 P. Fulde, H. Wagner: Phys. Rev. Lett. 27, 1280 (1971)
1.4 S. Takeno, M. Goda: Prog. Theor. Phys. 48, 1468 (1972)
1.5 P.W. Anderson, B.I. Halperin, C.M. Varma: Philos. Mag. 25, 1 (1972)
1.6 W.A. Phillips: J. Low. Temp. Phys. 7, 351 (1972)
1.7 H.B. Rosenstock: J. Non-Cryst. Solids 7, 123 (1972)
1.8 H.P. Baltes: Solid State Commun. 13, 225 (1973)
1.9 L.S. Kothari, Usha: J. Non-Cryst. Solids 15, 347 (1974)
1.10 G.J. Morgan, D. Smith: J. Phys. C7, 649 (1974)
1.11 D. Walton: Solid State Commun. 14, 335 (1974)
1.12 M.S. Lu: Ph. D. Thesis, Cornell University (1975)
1.13 P.R. Couchman, R.L. Reynolds, R.M.J. Cotterill: Nature 264, 534 (1976)
1.14 W.H. Tanttilla: Phys. Rev. Lett. 39, 554 (1977)
1.15 S. Hunklinger, W. Arnold, S. Stein, R. Nava, K. Dransfeld: Phys. Lett. A42, 253 (1972)
1.16 B. Golding, J.E. Graebner, B.I. Halperin, R.J. Schutz: Phys. Rev. Lett. 30, 223 (1976)
1.17 V. Narayanamurti, R.O. Pohl: Rev. Mod. Phys. 42, 201 (1970)
1.18 P.W. Anderson: Les Houches Lectures (North-Holland, Amsterdam 1978)
1.19 W. Arnold, S. Hunklinger: Solid State Commun. 17, 883 (1975)
1.20 B. Golding, J.E. Graebner: Phys. Rev. Lett. 37, 852 (1976)
1.21 A.P. Jeapes, A.J. Leadbetter, C.G. Waterfield, K.E. Wycherley: Philos. Mag. 29, 803 (1974)
1.22 J.R. Matey, A.C. Anderson: J. Non-Cryst. Solids 23, 129 (1977)
1.23 R.B. Stephens: Phys. Rev. B13, 852 (1976)
1.24 M. von Schickfus, S. Hunklinger: J. Phys. C9, L439 (1976)
1.25 E. Merzbacher: Quantum Mechanics, 2nd ed. (Wiley, New York 1970)
1.26 M. Abramowitz, I.A. Stegun: Handbook of Mathematical Functions (Dover, New York 1970)
1.27 M.E. Baur, W.R. Salzman: Phys. Rev. 151, 710 (1966)
1.28 J.A. Sussman: J. Phys. Chem. Solids 28, 1643 (1967)
1.29 N. Thomas: Ph. D. Thesis, University of Cambridge (1979)
1.30 J. Jäckle: Z. Phys. 257, 212 (1972)
1.31 W. Heitler: Quantum Theory of Radiation, 3rd ed. (Oxford University, Oxford 1954)

2. The Vibrational Density of States of Amorphous Semiconductors

D. L. Weaire

With 9 Figures

2.1 The Vibrational Density of States

In the classic approach to the study of vibrational properties of crystalline solids, the density of states $g(\omega)$ is derived (either experimentally or theoretically or by a combination of the two) from the dispersion relations $\omega(\underline{k})$ which give vibrational normal mode frequencies as functions of a Bloch vector $\underline{k}$. Take, for example, the painstaking work of NELIN and NILSSON [2.1], which established $g(\omega)$ rather accurately for germanium, as shown in Fig.2.1. This involved the interpolation, by means of a semi-empirical force constant model, of $\omega(\underline{k})$ data obtained from neutron scattering measurements for a large number of k points, and integration over the Brillouin zone.

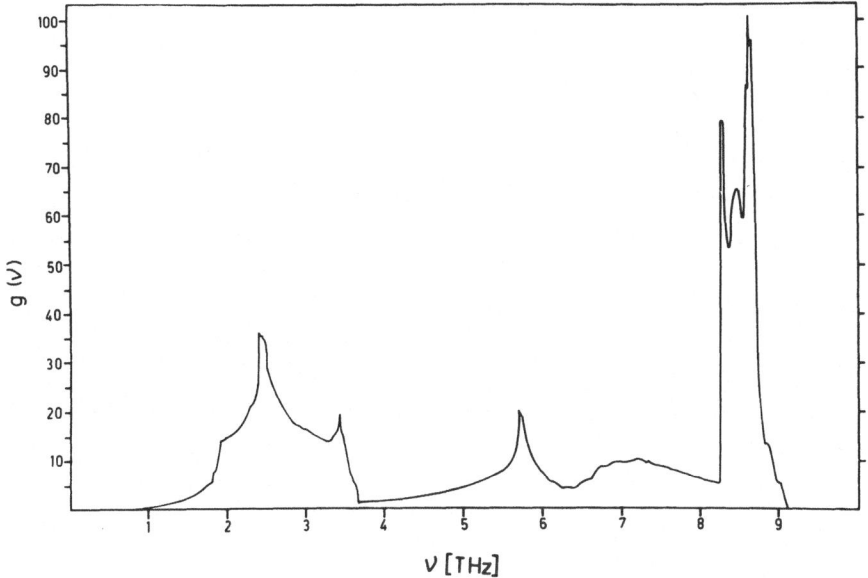

Fig.2.1. Vibrational density of states of crystalline Ge, as determined by NELIN and NILSSON [2.1]

For amorphous solids, the density of states $g(\omega)$ may be regarded as the primary quantity of interest, since there is no such thing as a Bloch vector in terms of which dispersion relations may be defined. How are we to determine $g(\omega)$ experimentally and theoretically?

The answer to the first part of the question has turned out to be rather simple — either infrared absorption or Raman scattering will give a good indication of the main features of the density of states. In other words, both of these processes detect *all* the normal modes, in sharp contrast to the selection of only (some) zone centre modes in crystals. Thus, the neutron scattering technique, which largely supplanted infrared absorption and Raman scattering in the case of crystals, plays a lesser role for amorphous solids.

At first, theoretical approaches to the problem were hesitant and depended largely on adapting ideas and calculations more appropriate to crystals. Since it turned out in the end that much is independent of periodicity, such efforts have tended to be superficially convincing but they have ultimately been unsatisfactory. Instead, one should avoid all mention of k vectors and related jargon and return to first principles.

2.2 Experimental Techniques

The purpose of this section is to give a general introduction to the three most useful spectroscopic techniques used in this field — IR absorption (or reflection), Raman scattering and neutron scattering. We shall not delve into the technical details of any of them.

The IR and Raman techniques have been widely used and routinely practiced by many laboratories. However, the interpretation of such spectra is by no means a matter of routine. Our remark in the opening section, to the effect that all vibrational modes are detected, must now be qualified by the observation that they do not appear with equal intensity. The transition probabilites, which determine the relative weightings of different modes in the spectrum, are not at all trivial. It is for this reason that it has been considered worthwhile to explore the application of inelastic neutron scattering (for which the transition probability *is* essentially trivial) to amorphous solids, despite the fact that this poses quite a challenge to existing experimental facilities.

In the case of *infra-red absorption*, the absorption coefficient is related to the density of states via the average of the square of the *dipole moment* associated with lattice vibrations of frequency ω (normalised to unit mean square amplitude). It is often written

$$\alpha(\omega) \sim |M(\omega)|^2 g(\omega) \qquad (2.1)$$

where M, the matrix element, is here to be understood in the sense of the average defined above. Occasionally, factors of ω^n are incorporated in the reduction of data with the claim that a better representation of $g(\omega)$ is thereby obtained. Since this is usually based on rather vague notions concerning the frequency dependence of M, confusion abounds. Only in the extreme low frequency regime can one hope to find a precise simple dependence.

The so-called *reduced Raman spectrum*, obtained from the raw data by the removal of various uncontroversial factors, also yields a weighted density of states. The contribution of a given mode in this case is determined by the squared modulus of the *polarisability* associated with the mode. This is a second rank tensor, to be combined with the polarisation vectors of incident and scattered light. The ratio of the Raman intensity for two different polarisations of the scattered light is (when suitably defined) termed the *depolarisation ratio* and it has been argued that this is a particularly significant quantity in the application of the technique to amorphous solids.

Turning to neutron scattering and again discarding relatively trivial factors, we find that the weighting of a mode n in the energy loss spectrum, for fixed scattering vector $\underset{\sim}{Q}$, is essentially determined by

$$I_n(Q) = |\sum_j \underset{\sim}{u}_j^n \cdot \underset{\sim}{Q} \, e^{i\underset{\sim}{Q}\cdot\underset{\sim}{R}_j}|^2 . \qquad (2.2)$$

Here $\underset{\sim}{R}_j$ denotes the position of the jth atom and $\underset{\sim}{u}_j^n$ is its displacement vector for the given mode n. The evaluation of (2.2) for a specified mode of a specified structure is straightforward. An additional advantage is gained if Q is given a very high value (relative to the inverse of the interatomic spacing). One may thus attain the *high Q limit*, in which the weighting function in (2.2) is approximately constant, independent of ω. It turns out, however, that this is not easily done and that there are other difficulties such as that presented by the contribution of multi-photon processes, which increases in this limit.

Results from IR and Raman experiments are shown in Figs.2.2 and 2.3. The IR results are the more uncertain. If the results are compared, it will be seen that the modes in the centre of the spectrum are the most strongly weighted in the IR spectrum, while the relative intensity appears to rise monotonically with frequency in the (reduced) Raman spectrum.

2.3 The Theoretical Problem

Two ingredients must be combined to provide a convincing analysis of the experimental data — a specific *structure* and a reasonable description of interatomic *forces*.

A variety of random network models are available for amorphous solids of current interest such as Si, Ge, SiO_2 and As. In each of these cases short range order is

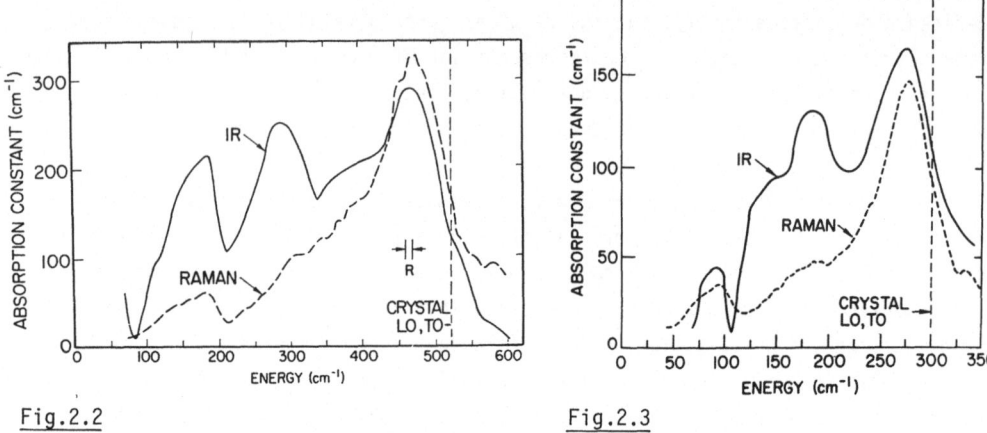

Fig.2.2

Fig.2.3

Fig.2.2. Infrared absorption coefficient (full line) and reduced Raman spectrum (dashed line) of amorphous Si [2.2,3]

Fig.2.3. Infrared absorption coefficient (full line) and reduced Raman spectrum (dashed line) of amorphous Ge [2.2,4]

combined with a bewildering variety of intermediate range configurations (usually discussed in terms of rings of bonds). We must not let the complexity of these distract us from the simplicity of the short range order, upon which much depends.

Equally, the intermediate forces are undoubtedly quite complicated, yet *nearest neighbour* central forces usually dictate the general form of $g(\omega)$. Figure 2.4, which is to be compared with Fig.2.1, illustrates this. The addition of simple non-central forces confined to nearest or, at most, second nearest neighbours provides an adequate description for most purposes.

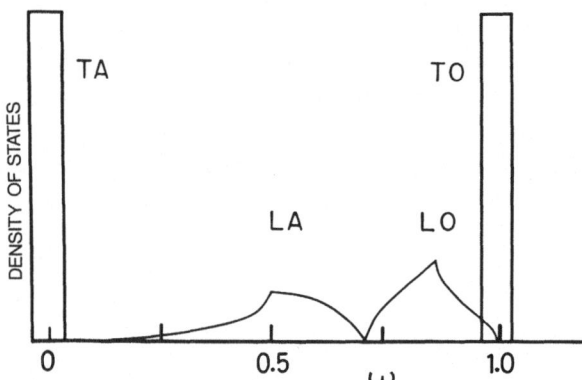

Fig.2.4. Vibrational density of states for the diamond cubic structure with central nearest neighbour forces [2.12]. (The highest frequency of the spectrum is here set equal to unity—as in Figs.2.5,7,8,9.) Delta functions are found at $\omega = 0,1$

2.4 Brute Force Theory

2.4.1 Turning the Handle

Brute force, by which we mean the direct calculation of vibrational densities of states for specific structural models, has much to recommend it. What emerges is a straightforward test of the assumptions upon which the calculation is based, unobscured by speculative approximations and other theoretical sleight-of-hand. But one should not try to set up a dichotomy between this and the more elegant approaches of Sect.2.5 — they are clearly complementary and interdependent.

Given a model structure all we have to do is to define suitable interatomic forces and "turn the handle" of a standard normal mode calculation for a finite cluster. The only difficulties are

a) It may not be easy to diagonalise the secular determinant for a sufficiently large cluster for it to be representative of an infinite system.

b) The *surface* will seriously distort the spectrum unless careful precautions are taken.

c) Interatomic forces may be subject to some uncertainty.

The first people to confront these problems were DEAN [2.5] and BELL [2.6], in studying SiO_2, and their response to the challenge was as follows:

a) They did not use the standard matrix diagonalisation routines of the time, but instead used the *Sturm sequence* method, which was comparatively efficient in yielding a density of states for large systems, so that they could study clusters of $N \sim 300$ atoms.

b) They used both *free* and *fixed* boundaries, so that spurious features due to the surface could be distinguished.

c) They used only short range forces of simple form.

The subsequent developments in each of these three categories can be summarized as follows:

a) The most efficient method which follows this general (i.e., fairly *direct*) approach is the Lanzcos method [2.7], by means of which $N \sim 10^3$ atoms can be handled. Alternatively, less direct methods can be used for systems of similar size, as discussed in Sect.2.5.

b) There is one celebrated random network model [2.8] with *periodic* boundary conditions (i.e., the model can be periodically continued with appropriate bond-ing configuration at the boundaries).
More generally, one may use special boundary conditions (Sect.2.5.2) or projection to eliminate most surface effects.

c) There is still much uncertainty if forces of longer range than nearest neighbours are introduced. Most work still uses very simple short range forces (see,
however, Sect.2.7).

2.4.2 Some Results of Brute Force

For the group IV semiconductors, the Born model [2.9] provides a simple two parameter force constant scheme for bond stretching and bending forces. For a given
set of atomic displacements $\underset{\sim}{u}_i$ of atoms with equilibrium positions $\underset{\sim}{r}_i$, each bond
contributes to the potential energy a term,

$$\frac{3\beta}{2}\,[(\underset{\sim}{u}_i - \underset{\sim}{u}_j)\cdot\hat{r}_{ij}]^2 + \frac{\alpha-\beta}{2}\,|\underset{\sim}{u}_i - \underset{\sim}{u}_j|^2 \qquad (2.2)$$

where $\hat{r}_{ij}$ is a unit vector parallel to $\underset{\sim}{r}_i - \underset{\sim}{r}_j$.
   Alternatively, if we write $u_{||}$ and $u_{\perp}$ for the magnitude of the components of
the relative displacement $\underset{\sim}{u}_i - \underset{\sim}{u}_j$ parallel and perpendicular to the bond vector
$\underset{\sim}{r}_{ij}$, this may be written

$$\left(\frac{\alpha+\beta}{2}\right)u_{||}^2 + \left(\frac{\alpha-\beta}{2}\right)u_{\perp}^2 \quad . \qquad (2.3)$$

The first term represents the central or bond stretching force. Figure 2.5 shows
the calculated [2.10] density of vibrational states for the diamond cubic structure with bond bending forces of a relative strength appropriate to Si or Ge. This
reproduces the main feature of Fig.2.1 (cf. also Fig.2.4). Born forces or keating
forces [2.9], which are slightly different, therefore offer a reasonable basis for
a calculation of $g(\omega)$ for *amorphous* Si or Ge, provided that we do not look at the
very finest details of the spectrum. Figure 2.6 shows the results of such a calculation, performed by ALBEN et al. [2.11] for the Henderson model (with periodic
boundary conditions). Although individual eigenvalues were calculated, the results
were presented as à sum of narrow Lorentzian contributions from these, as an alternative to a histogram.

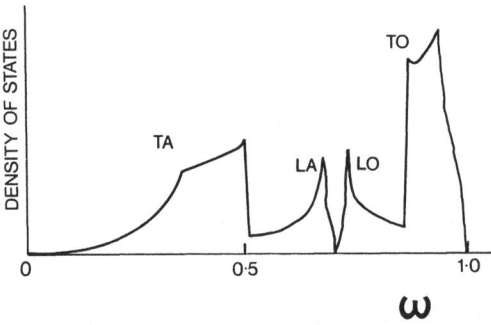

Fig.2.5. Vibrational density of states for the diamond cubic structure with Born
forces of relative strength appropriate to Si or Ge [2.10]

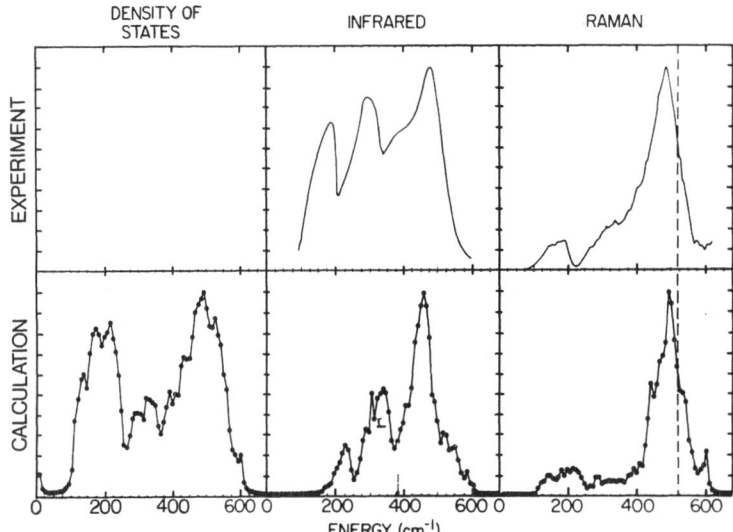

Fig.2.6. "Brute force" calculation of the vibrational density of states and IR and Raman spectra of a random network model of amorphous Si, with Keating forces [2.11]

What emerges from this quite substantial calculation is a density of states not unlike that of the crystal (Figs.2.1 and 2.5). This is a conclusion entirely con- sistent with experiment, and is important despite its rather negative nature, for it says that the maintenance of the same broad features in the amorphous state is *not* evidence for the microcrystalline model. It is quite consistent with the random network model.

We must look at the details of these results if we are to discern significant features, but must not overdo this, as explained above. One immediately apparent difference between Figs.2.1 and 2.6 lies in the width of the upper ("TO") peak. Suffice it to say here that the broadening of this peak is again consistent with experiment.

Yet finer details have been discussed from time to time. Has the lowest ("TA") peak moved? Can one discern the "LA" and/or "LO" peaks in the middle? These matters remain largely unsettled.

To summarise, a straightforward "brute force" calculation is not difficult, but its finer details tend to be controversial. Of great help at this point are the more powerful numerical methods and the simple analytical ideas of the next section.

2.5 More Refined Approaches

2.5.1 Analytical Ideas

Playing the role of Beauty alongside the Beast of computer calculations are a number of analytical theorems.

Let us first return to Fig.2.4 to see how the structure of $g(\omega)$ may be understood in general terms, not relying on language appropriate only to the diamond cubic structure. Note first that the "TA" peak at $\omega = 0$ is derivable from the following general considerations. For a system of N atoms with 3N degrees of freedom there are 3N - 2N = N ways of displacing the atoms if 2N bonds are to remain of fixed length. These correspond to N zero frequency modes for central forces, i.e., one per atom. Note that, in general, the label "TA" is *not* justifiable but even purists such as the present author tend to use it!

What about the rest of the spectrum? A theorem of WEAIRE and ALBEN [2.12] states that for a perfectly tetrahedrally coordinated structure and central forces, there is another such delta function at the top of the spectrum, as in Fig.2.4, again independent of structure. Between the two stretches the remaining one-third of the density of states which, on an ω^2 scale, is isomorphic with that of the simple s-band tight-binding Hamiltonian

$$H = \sum_{ij} V|i><j| \qquad (2.4)$$

coupling nearest neighbours only. This, of course, is somewhat dependent on structure.

Note also that we used *perfect* tetrahedral coordination only to derive the *upper* delta function so it should presumably be broadened by local distortions *even in the central force model*. This was demonstrated in brute force calculations by WEAIRE and ALBEN [2.12]. In this way the broadening of the upper peak noted in Sect.2.4 is explained and associated with a specific structural feature.

We have excluded the bond bending forces, which are clearly necessary if we are to further interpret the calculations. There is not much that can be proved in any exact sense once these are included.

2.5.2 The Bethe Lattice

The Bethe lattice provides an alternative way of seeing that the broad features of the density of states arise from tetrahedral coordination alone.

A variety of approximate physical theories of the "mean field" type can be regarded as exact solutions for this artificial *pseudo* lattice. It is simply an infinite branching structure, in which each vertex (atom) has the same nearest neighbour shell, a tetrahedral arrangement in the case of most interest to us.

It is amenable to very straightforward solution, by a variety of methods, if
the interactions (forces) are of short range. The introduction of longer range
forces complicates matters greatly.

The artificial nature of the structure is manifested in the fact that it has
a density of states whose band edges fall short of the values attained in real
lattices. This is particularly embarrassing for lattice vibrations since it entails
a gap between $\omega = 0$ and the lower band edge.

A calculation [2.13] of $g(\omega)$ for the Bethe lattice appropriate to tetrahedrally
bonded Si or Ge is shown in Fig.2.7.

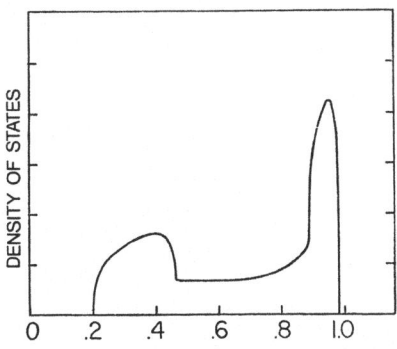

Fig.2.7. Vibrational density of states for a
Bethe lattice, with local tetrahedral coordination and Born forces of relative strength
approximate to Si and Ge [2.13]

2.5.3 Indirect Numerical Methods

There are two important methods for the determination of $g(\omega)$ for model structures
which are less direct than matrix diagonalisation and allied methods. These are the
recursion method [2.14] and the *equation-of-motion method* [2.15]. They are, of
course, of much wider applicability than is considered here. We shall try to indicate the spirit of these methods rather than their details.

The recursion method forces the problem into the form of a *linear chain*, so let
us first look at a Hamiltonian of this form, illustrated below

We define elements of an Hermitian Hamiltonian according to

$$\langle i|H|i\rangle = a_i$$
$$\langle i|H|i-1\rangle = b_i$$
$$\langle i|H|j\rangle = 0 \text{ for } i \neq j, j \pm 1 \ . \tag{2.5}$$

The Green's function (whose imaginary part for E just above the real axis gives
the density of states), which is defined by

$$G = (E - H)^{-1} \ , \tag{2.6}$$

has the matrix element

$$G_{00} = <0|G|0> = \cfrac{1}{E - a_0 - \cfrac{|b_1|^2}{E - a_1 - \cfrac{|b_2|^2}{\cdots}}}. \qquad (2.7)$$

This may be easily proved by invoking Löwdin perturbation theory [2.16], which allows us to lop off the last link in the chain, provided that we modify the diagonal matrix element to the Hamiltonian, associated with the $(n - 1)^{th}$ site, making it

$$a_{n-1}^{(eff)} = a_{n-1} - \frac{|b_n|^2}{E - a_n}. \qquad (2.8)$$

The iterative application of this procedure allows us to lop off all links of the chain except the first, building up a continued fraction associated with the last link, until it is the only one left, at which stage (2.7) follows. Somewhat intuitively, we may assume that the *infinite* continued fraction is equally representative of semi-infinite chain.

Now, in a three dimensional solid, we may focus on any coordinate (or combination of coordinates) to play the role of $|0>$ above and generate further combinations of coordinates to form the basis set such that the dynamical matrix D takes the form of a semi-infinite chain. To do so, one solves

$$b_{n+1}|n + 1> = (D - a_n I)|n> - b_n^*|n - 1> \qquad (2.9)$$

recursively, together with the requirement of orthonormality for the new basis set. Then, with the translation $E \Leftrightarrow \omega^2$, the above formula (2.7) may be applied. Thus may the complexities of any given Hamiltonian be stuffed into the sausage of linear chain form! In practice, one must terminate the recursion at some point, and the manner in which this is done is the most subtle point of the method, which is otherwise straightforward and extremely economical in computer time and storage.

The total density of states $g(\omega)$, can of course, be obtained by summing the local density of states associated with each individual coordinate (see also the remarks at the end of the section).

Some results of MEEK [2.10], obtained by this method are shown in Fig.2.8.

The equation-of-motion method [2.15] has similar virtues but proceeds rather differently. Let us again choose some coordinate and calculate the time dependence of a vector of displacements, whose starting value, denoted by $|0>$, is zero, apart from the single coordinate which we may choose to give the value unity. Now, if the vibrational eigenstates are denoted by $|n>$, we may write

$$|0> = \sum_n |n><n|0> . \qquad (2.10)$$

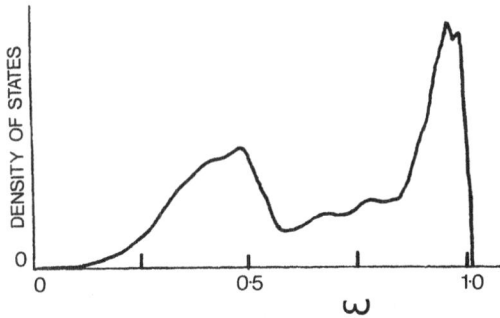

DENSITY OF STATES

Fig.2.8. Vibrational density of states for a random network model of amorphous Si, as calculated using the recursion method, with Born forces [2.10]

Hence the time dependence of the coordinate in question will just be

$$\langle 0|t\rangle = \sum_n \langle 0|n\rangle \langle n|0\rangle \cos\omega_n t \quad . \qquad (2.11)$$

The determination of $\langle 0|t\rangle$ by the integration of the equation of motion and a subsequent Fourier transform will yield the quantity

$$\sum_n \delta(\omega - \omega_n)|\langle n|0\rangle|^2 \qquad (2.12)$$

which is just the local density of states associated with the coordinate in question.

There are numerous subtleties [2.15] in the method as it is used in practice. In particular, starting vectors consisting of random choices for *all* coordinates may be used to generate the *total* density of states directly. (This can also be done in the recursion method).

Some results of BEEMAN and ALBEN [2.15] obtained by this method are shown in Fig.2.9, providing an interesting comparison with the previous figure.

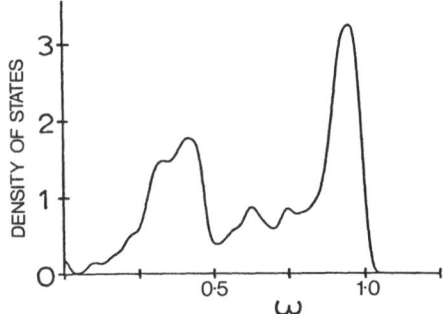

Fig.2.9. Vibrational density of states for a random network model of amorphous Si, as calculated by the equation of motion method, with Born forces [2.15]

Both of these methods can be further adapted [2.15] to introduce, by a suitable choice of starting vector, a weighting appropriate to a given choice of the matrix elements determining infra-red or Raman transition probabilities (Sect.2.2).

Both methods can also straigthforwardly project out the local density of states associated with only the central part of a large cluster, thereby mitigating the problem of choice of boundary conditions.

2.6 The Incorporation of Matrix Elements

We have already explained (in Sect.2.2) the importance of the transition probabilities in the IR and Raman spectra, and seen their effects reflected in the difference between the spectra of Fig.2.3.

For the IR spectrum, the usual simple first approximation to $\underline{M}$ is derived by associating *point* charge with the atoms, different for each species. In the homopolar case something more subtle is required and ALBEN et al. [2.11] chose to write $\underline{M}$ as a sum of local contributions, as follows.

A transfer of charge between adjacent bonds was assumed, its magnitude depending on the relative compression or extension of the bonds. In as much as this simple *ad hoc* mechanism is quite successful it is worth considering why. The fact that the (root mean square) average compression of bonds varies as ω in the central force approximation might lead one to expect an ω^2 dependence from such a mechanism. However, cancellation associated with tetrahedral symmetry suppress the upper part of the spectrum, as shown by ALBEN et al. [2.11].

As for the Raman spectrum, the use of a similar semi-empirical form (but rather more involved since the Raman intensity is given by a *tensor*) can also give reasonable agreement with experiment (Fig.2.4). Here one has a frequency dependence which is *approximately* ω^2, again arising from the average compression of bonds.

2.7 Can One Derive Structural Information from g(ω)?

Can the process by which we have come to understand the shape of $g(\omega)$ in terms of the random network model for Si and Ge be *inverted* to analyse the structure itself? Various tentative suggestions have been made concerning this. Certainly some significant variation of $g(\omega)$ seemed to occur in calculations [2.10,11] when Born or similar forces were used. The prospect emerges of at least a rough estimate of the *ring statistics* of the network.

Unfortunately, MEEK [2.17] has demonstrated that the effects of longer range interactions, as given by the Weber bond charge model [2.18], is to wash out most of the variable features in the earlier calculations. For the present, at any rate, this point of view holds sway.

2.8 A Less Myopic View of the Field

Our single-minded preoccupation with Si and Ge up to this point needs to be balanced by a wider view of the field.

Much of what we have said applies rather generally. In most amorphous solids, the density of states which, with some distortion, is seen in Raman and infrared absorption spectra, is dominated by the effects of short range forces and nearest neighbour coordination and hence matches the density of states of crystalline solids with similar short range order. This is not to say, however, that the dominant features of the density of states are always *trivially* related to the short range order, as in the case of molecular solids. The "molecular" model, in which the spectrum is interpreted in terms of the modes of a *small* isolated cluster, must be used with caution [2.19]. One way of making it more reliable is to use special boundary conditions. For example, one may use effective potentials which are equivalent to attaching branches of the Bethe lattice to the surface bonds. This is the so-called "cluster-bethe" model [2.20].

In general, just as we saw for the case of Si and Ge, an analysis of the vibrational density of states supports the random network model without providing a sensitive probe of its finer details. There is one very important exception to be found in the recent work of BRODSKY et al. [2.21] on glow discharge deposited Si. Raman and infrared spectra provide a measure of the quantity of hydrogen present and the relative fractions of hydrogen bonded singly, in pairs, etc. to the Si network. A more comprehensive survey of such matters has been made by WEAIRE and TAYLOR [2.22]. Some of the problems which remain for study in the field are as follows:

a) Can one have something analogous to LO-TO splittings in amorphous solids? If so, how are they to be calculated? This question has been raised by GALEENER and LUCOVSKY [2.23].

b) Can a feasible experiment be devised, by means of which *localised* vibrational eigenstates are distinguishable from *extended* ones? DEAN [2.5] and BELL [2.6] made much of this distinction but it has not proved relevant to experiment so far.

c) What are the asymptotic forms of Raman and infrared spectra at low frequency? The literature is littered with suspect reasoning on this matter. For example, PRETTL et al. [2.24] suggested $\alpha \sim \omega^4 n(\omega)$, on the basis of an expansion of M [in (2.1)] in powers of a wave vector q. But q is not a good quantum number. (If it were, all $q \neq 0$ would contribute *zero*!)

References

2.1 G. Nelin, G. Nillson: Phys. Rev. B*5*, 3151 (1972)
2.2 M.H. Brodsky, A. Lurio: Phys. Rev. B*9*, 1646 (1974)
2.3 J.E. Smith, Jr., M.H. Brodsky, B.L. Crowder, M.I. Nathan, A. Pincuk: Phys. Rev. Lett. *26*, 642 (1971)
2.4 J.E. Smith, Jr., M.H. Brodsky, B.L. Crowder, M.I. Nathan: In *Proc. 2nd Int. Conf. on Light Scattering in Solids*, ed. by M. Balkanski (Flammarion, Paris 1971) p.330
2.5 P. Dean: Rev. Mod. Phys. *44*, 127 (1972)
2.6 R.J. Bell: Rep. Prog. Phys. *35*, 1315 (1972)
2.7 J.T. Edwards, D.J. Thouless: J. Phys. C*5*, 807 (1972)
2.8 D. Henderson: In *Computational Solid State Physics*, ed. by F. Herman (Plenum, New York 1972) p.175
2.9 See, e.g., R.M. Martin: Phys. Rev. *1*, 4005 (1970)
2.10 P.E. Meek: Philos. Mag. *33*, 897 (1976)
2.11 R. Alben, D. Weaire, J.E. Smith, Jr., M.H. Brodsky: Phys. Rev. *11*, 2271 (1975)
2.12 D. Weaire, R. Alben: Phys. Rev. Lett. *29*, 1505 (1972)
2.13 M.F. Thorpe: In *Amorphous and Liquid Semiconductors*, ed. by J. Stuke and W. Brenig (Taylor and Francis, London 1974) p.835
2.14 R. Haydock, V. Heine, M.J. Kelly: J. Phys. C*5*, 2845 (1972); *8*, 2591 (1975)
2.15 D. Beeman, R. Alben: Adv. Phys. *26*, 339 (1977)
2.16 See, e.g., A.R. Williams, D. Weaire: J. Phys. C*9*, L47 (1976)
2.17 P.E. Meek: Thesis, University of Cambridge (1977)
2.18 W. Weber: Phys. Rev. Lett. *33*, 371 (1974)
2.19 P.N. Sen, M.F. Thorpe: Phys. Rev. *15*, 4030 (1977)
2.20 J.D. Joannopoulos, F. Yndurain: Phys. Rev. *10*, 5154 (1974)
2.21 M.H. Brodsky, M. Cardona, J.J. Cuomo: Phys. Rev. B*16*, 3556 (1977)
2.22 D. Weaire, P.C. Taylor: In *Dynamical Properties of Solids*, Vol.4, ed. G.K. Horton, A.A. Maradudin (North-Holland 1980) p.1
2.23 F.L. Galeener, G. Lucovsky: Phys. Rev. Lett. *37*, 1471 (1976)
2.24 W. Prettl, N.J. Shevchik, M. Cardona: Phys. Status Solidi B*59*, 241 (1973)

3. Low Temperature Specific Heat of Glasses*

R. O. Pohl

With 17 Figures

One of the puzzles presented by amorphous solids is that their low-temperature specific heat exceeds that predicted on the basis of the Debye theory by a considerable amount. This phenomenon appears to be sufficiently general to be considered a
characteristic one. However, it is not altogether clear what fraction, if any, of
the specific heat anomaly is truly intrinsic, i.e., independent of the prior history
of the sample and of the presence of trace impurities. In this chapter we will review the experimental situation.

3.1 Review of the Experimental Situation

One of the major achievements of the theory of lattice dynamics was the recognition
that the atomic vibrations of solids can be described by plane waves in the continuum approximation [3.1]. The Debye model of the lattice specific heat of solids
has been very successful in describing the experimental observations in chemically
pure crystalline solids. As examples Fig.3.1 shows the specific heats of KCl, TiO_2
and graphite [3.2-8]. The solid curves were computed with the three-dimensional
Debye model on the basis of measured speeds of sound. Note the T^3 temperature dependence at low temperatures. The model fits the KCl data very well and the TiO_2
reasonably well. The origin of the discrepancy between theory and experiment is
known to be the dispersion of the lattice vibrations in the crystals resulting from
the discontinuity of the solid on the atomic scale. The humps visible in TiO_2
around $T = 0.1\,\theta$ (θ is the Debye temperature), and to a lesser degree also in KCl,
are caused by the higher density of states of the lowest transverse acoustic
branch near the edge of the first Brillouin zone, as observed directly through
neutron scattering. At temperatures $T > 10^{-2}\,\theta$, the specific heat of graphite varies
as T^2. This can be understood through the two-dimensional layer structure of this
carbon allotrope. Note that even this substance displays a lattice specific heat
varying as T^3 at very low temperatures.

*This work was partly supported by the U.S. National Science Foundation, Grant
No. DMR 78-01560.

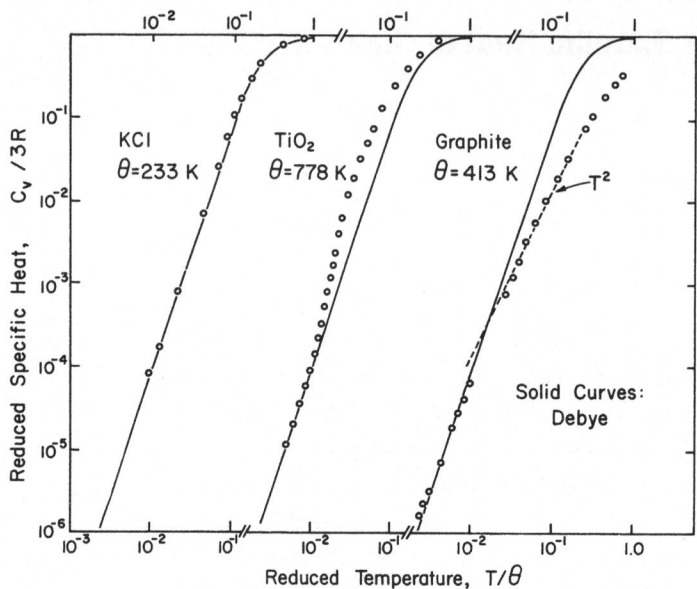

<u>Fig.3.1.</u> Reduced specific heat for three different crystals, compared with the Debye theory based on elastic measurements (solid lines). R - Gas Constant = 8.31 W s mole^{-1} K^{-1}, mole = (A) gram, with (A) being the average atomic weight of the host *atoms*. θ = Debye temperature. KCl — [3.2,3]; TiO$_2$ — [3.4,5]; Graphite — [3.6-8]

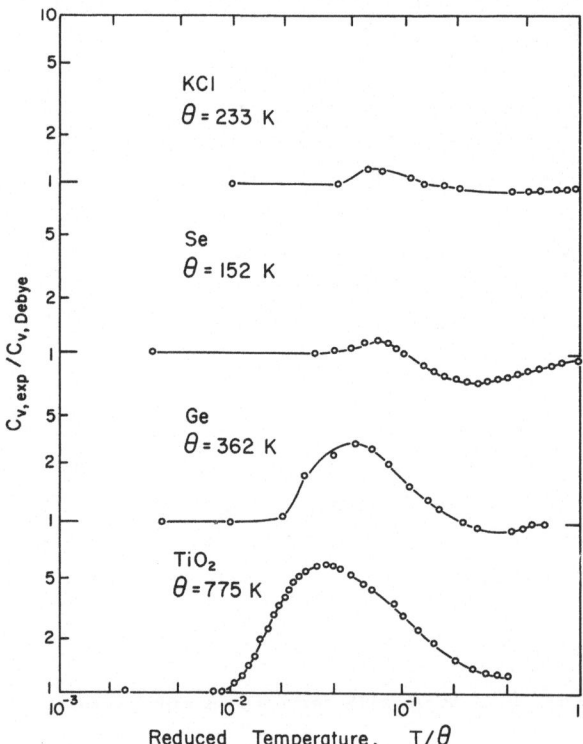

<u>Fig.3.2.</u> Deviation of the experimental specific heat $C_{v,exp}$ from the Debye prediction, plotted as $C_{v,exp}/C_{v,Debye}$. KCl and TiO$_2$ — see Fig.1; Se — [3.9]; Ge — [3.10,11]

The quality of the fit using the Debye model is shown more clearly in Fig.3.2, where the ratio of the measured specific heat to that calculated with the Debye model using only the measured sound velocities has been plotted for four crystals [3.9-11]. In all pure crystals studied to date, the measured specific heat agrees with the Debye prediction at temperatures below $10^{-2}\theta$ to within a few percent. It seems reasonable to conclude that in crystalline solids at least, dispersion has no influence on the plane wave excitations whose wavelengths exceed $\sim$100 lattice spacings.

In amorphous solids, one would expect the continuum approximation underlying the Debye model to apply at least as well as in crystalline solids. The experiment, however, have shown the low-temperature specific heat to be very different from the Debye prediction. This is shown in Fig.3.3 for three solids which exist in the crystalline as well as in the amorphous phase [3.12-15]. The hump characteristic for the dispersion of the transverse acoustic phonons occurs at a lower tempera-ture, in agreement with neutron scattering results (e.g., for SiO_2 see [3.16]). In both amorphous SiO_2 and polyethylene (PE), however, the Debye T^3 regime has not yet been reached at $10^{-2}\theta$. In Se, in fact, the ratio C_{exp}/C_{Deb} increases as the temperature is lowered. This behavior at temperatures below $10^{-2}\theta$ is probably com-mon to all noncrystalline solids. Figure 3.4 shows one of the first observations of this phenomenon [3.9]. The specific heat of the glassy phase decreases much more slowly than the Debye T^3 prediction; the dashed line indicates this predic-tion (which is almost the same for the crystalline and the amorphous phase for this substance). The specific heat of amorphous SiO_2 is well described by an expression of the form

$$C = c_1 T + c_3 T^3 \quad , \quad 0.1K < T < 1K \quad , \tag{3.1}$$

where c_3 exceeds the Debye prediction. The experiments performed and reviewed by ZELLER and POHL and later by STEPHENS [3.17] indicated that the specific heat was fairly independent of the silica sample (see, e.g. [Ref.3.9, Fig.13] and [Ref.3.17, Fig.4]). The work by LASJAUNIAS et al. [3.18] which was extended to 0.025K, how-ever, clearly showed the approximate nature of (3.1) (confirmed in [3.18a]), and also demonstrated the extent of the sample dependence of the anomaly (see Fig.3.5). LASJAUNIAS et al. [3.18b] have suggested that the steeper temperature dependence resulted from a gap in the low energy density of states; for SiO_2 (Suprasil W), the magnitude of the gap was estimated as having width $E_{min} = 16mK \cdot k_B$.

Although the larger specific heat at the lowest temperatures is found in a sample whose OH contents are higher, it is not possible to identify these impurities with all or at least a part of the anomaly, since no systematic study of the in-fluence of this impurity has been made. In fact, the only case in which an impurity in a glass has been clearly identified through low-temperature specific heat measure-ments is iron in borosilicate glass [3.9,17,19], as shown in Fig.3.6. The presence of iron [100 ppm by weight in the Corning No. 7740 glass ("Pyrex"), and 12 ppm in

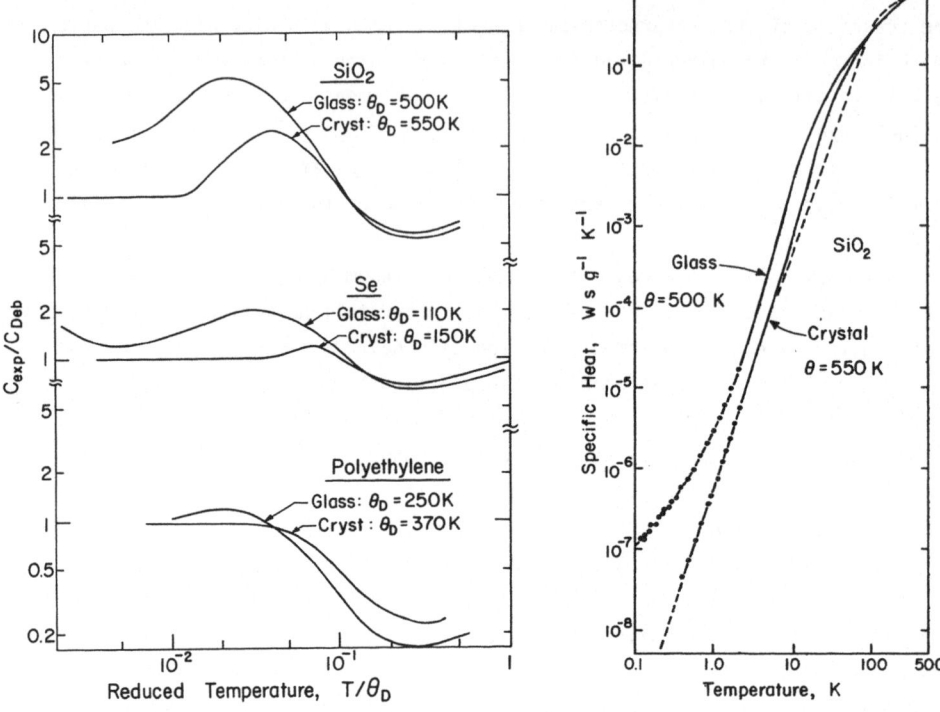

Fig.3.3   Fig.3.4

Fig.3.3. Comparison of the experimental specific heat of SiO$_2$, Se, and polyethylene (PE) in their glassy and crystalline phases, plotted as C_{exp}/C_{Deb} vs T/θ (after [3.12]). The crystal phase of SiO$_2$ is α-quartz. θ for the two phases of SiO$_2$ was determined from speed of sound measurements reviewed in [3.9]; θ for amorphous Se from ultrasonic measurements in [3.13]; and for crystalline Se from specific heat measurements after [3.14]. θ for crystalline PE was determined from specific heat measurements in [3.15]. For amorphous PE the same procedure was used; this is some-what dubious, as discussed in this Chapter. However, a shift of the specific heat curve for amorphous PE along the T-axis does not alter the point which this figure is supposed to make

Fig.3.4. Specific heat of SiO$_2$ in its amorphous and crystalline (α-quartz) phases. The large specific heat anomaly varies almost as T at low temperatures (after [3.9]). Note that this low temperature anomaly is somewhat sample dependent, as shown in Fig.3.5

the No. 9700 glass, according to chemical analysis] results in a specific heat anomaly which peaks at around 0.5K. In a 33kG magnetic field, the specific heat observed in the 9700 glass is very close to that of iron free soda-silica glass. From the difference, the iron-induced entropy can be determined, assuming two-level systems. These impurity concentrations, 220 and 50 ppm, are close enough to the iron concentration determined by chemical analysis to allow the conclusion that the magnetic field sensitive extra anomaly is caused by the iron trace impurities. Interestingly, however, the iron-induced anomaly clearly does not have a shape of the form of (3.1).

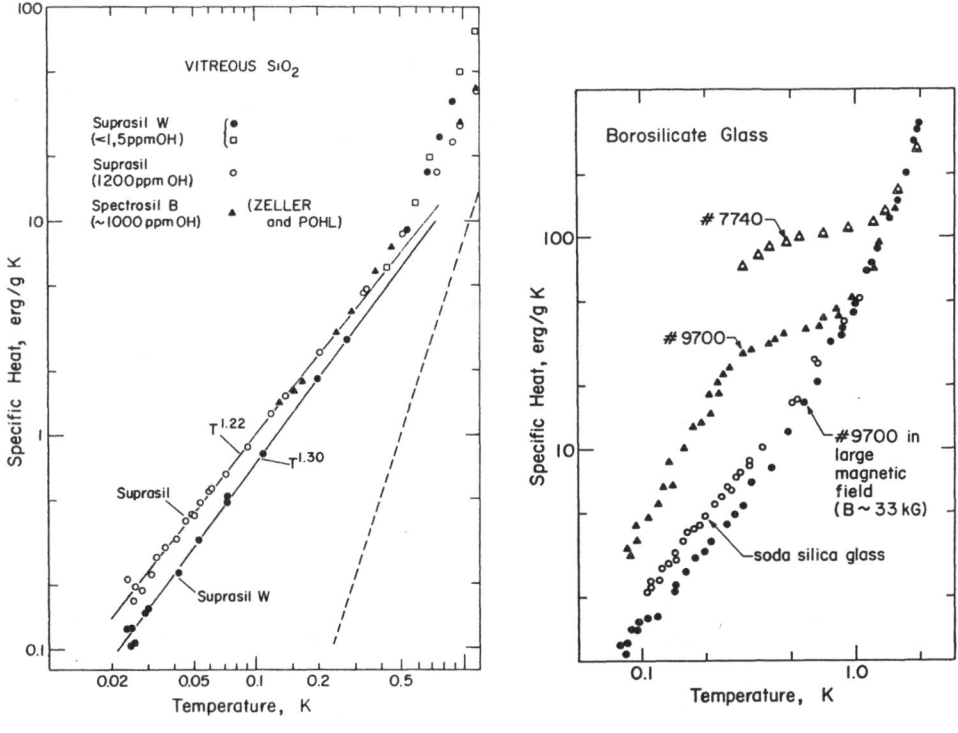

Fig.3.5

Fig.3.6

Fig.3.5. Specific heat of vitreous SiO₂ above 0.025K [3.18]. Suprasil and Spectrosil B have large OH concentrations, but small metal ion concentrations. Suprasil has 130 ppm chlorine and 100 ppm fluorine. Suprasil W has low OH and metal ion concentrations, but 230 ppm chlorine and 290 ppm fluorine (all numbers according to the manufacturers). Note that Spectrosil B and Suprasil have identical specific heats in the temperature range where both were measured (T > 0.1K). The dashed line marks the Debye specific heat of vitreous SiO₂, $C_{Deb} = 8\ T^3$ erg/(gramK⁴), based on $v_t = 3.75 \times 10^5$ cm/sec and $v_\ell = 5.80 \times 10^5$ cm/sec

Fig.3.6. Specific heat of Corning borosilicate glasses code No. 7740 and No. 9700. The No. 7740 data (Δ) after [3.9], all other data from [3.17]; the No. 9700 data were made in both 0-kG(▲) and 33-G(●) magnetic fields. Data from a soda-silica glass (3SiO₂,Na₂O) are included (o) for comparison to the borosilicate's 33-kG heat capacity. The sizes of the anomaly are roughly those expected from the measured concentration of iron in these glasses: 100 ppm for No. 7740, and 12 ppm for No. 9700

In an effort to detect the effect of thermal treatment on the specific heat of a glass, WENGER et al. [3.20] measured a sample of silica Suprasil glass (1200 ppm OH) before and after heating the sample in air at 1150°C, above the strain point (1125°C) for 1.5 hours, followed by a quench in an oil bath, which produced uniform strain, but no fissures. Within their experimental accuracy no influence of this heat treatment on the specific heat was found in the temperature range of their experiment (0.4K < T < 1.4K). The data were indistinguishable from those obtained by LASJAUNIAS et al. [3.18] on Suprasil and by ZELLER and POHL [3.9] on Spectrosil B (see Fig.3.5).

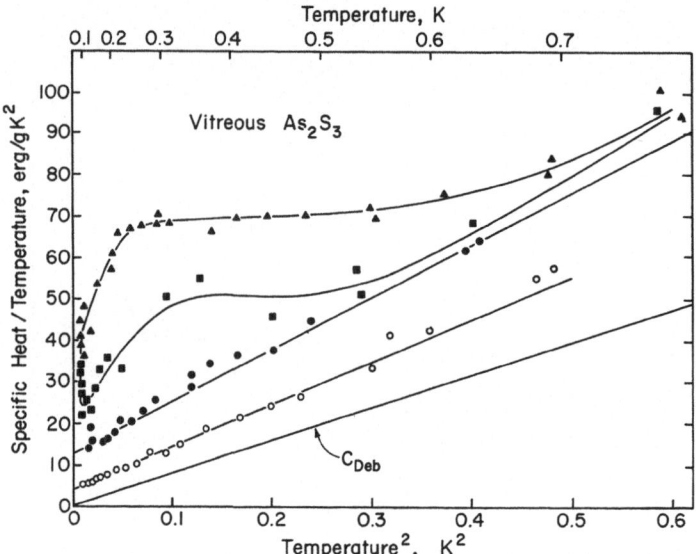

Fig.3.7. Specific heat of four As2S3 samples plotted as C/T vs. T^2 [3.17]. The line marked C_{Deb} is the heat capacity predicted by the Debye model. The two upper curves are from samples which were produced at Cornell with successively more care to remove water of hydration. The next curve is from a sample which was supplied by A.J. Leadbetter. The lowest set of data is from a sample which was made by F.J. DiSalvo at Bell Labs. If one assumes that the differences between the upper three and lowest sets of data are due to the presence of two-level systems, one can cal-culate the density of these systems to be 68×10^{16} cm^{-3} for the upper curve, 26×10^{16} cm^{-3} for the second one, and $\sim 10^{16}$ cm^{-3} for the third. Spark source mass spec analysis of the four measured samples showed the following impurity concen-trations in the order given above: 1) 10^{19} to 10^{20} cm^{-3} Sb and 10^{17} to 10^{18} cm^{-3} Rb; 2) 10^{17} to 10^{18} cm^{-3} and 10^{16} to 10^{17} cm^{-3} Cd; 3) 10^{17} to 10^{18} cm^{-3} Ge; 4) nothing detectable

Other glasses have similar anomalies. Figure 3.7 shows data (after [3.17]) for four different samples of As$_2$S$_3$ plotted as C/T vs T^2 on linear scales. In such a graph, a specific heat of the form of (3.1) will be represented by a straight line whose intercept with the vertical axis is c_1, and whose slope equals c_3. Apparently, purer samples have a lower specific heat (see figure caption of Fig.3.7), although no scaling between specific heat and impurity concentration has been established. In the purer samples, (3.1) produces a good fit to the data. Note the discrepancy between the measurements and the Debye-predicted specific heat shown in Fig.3.7; this figure raises the obvious question whether a more careful purification or a different sample preparation might further lower the specific heat. This question has been only partially answered through the work on Se, of which three different samples have been measured. One, the less carefully prepared sample, had a higher specific heat which could not be very well described with (3.1) [3.14]. The two other samples, one measured by LASJAUNIAS and THOULOUZE [3.21], the other by STEPHENS [3.22], were found to have identical specific heats (lower than the first sample), even though they had been prepared in different laboratories. Their specific heats could be fitted well with an expression of the form of (3.1) [3.22]. After the measurement, LASJAUNIAS et al. recrystallized their less carefully prepared sample [3.14]. In this state, its specific heat was close to that of the known specific heat of crystalline Se. This shows that whatever "impurities" of "imperfections" cause the glass anomaly in the glassy selenium, they are much less effective in the crystalline phase.

In B_2O_3, the larger specific heat of a less carefully dried sample was found to be well represented by an expression of the form of (3.1), while the lower specific heat of more water-free samples showed a deviation from this form below 0.2K [3.17]; in this range, LASJAUNIAS et al. [3.23] obtained a better fit with a single $T^{1.45}$ power law.

In the ionic glass $KCa(NO_3)_3$, STEPHENS [3.17] found a lower specific heat in less water containing, drier samples, while (3.1) produced only a moderately good description of the specific heat in all cases. Similarly, the specific heat of the polymers polystrene (PS) and polymethylmethacrylate (PMMA) and of a soda-silica glass can only be approximated with an expression of the form of (3.1), as shown in Fig.3.8 [3.17].

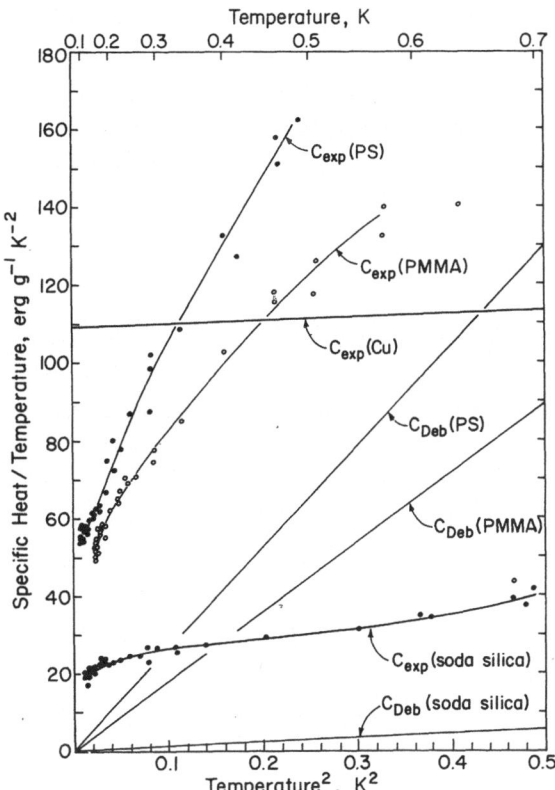

Fig.3.8. Specific heat of three noncrystalline solids plotted as C/T vs T^2: $3SiO_2 \cdot Na_2O$ (soda silica), Polystyrene (PS), and Polymethylmethacrylate (PMMA). The specific heats predicted by the Debye model, the lines marked C_{Deb}, are shown for each of the materials. The specific heat of Cu is shown for a comparison of its electronic contribution to the size of the excess $(C_{exp} - C_{Deb})$. [3.17]

In an epoxy resin, the experimental accuracy was inadequate to determine the temperature dependence accurately [3.23], however, the authors obtained a satisfactory fit to the data with an expression of the form of (3.1).

The only bulk amorphous electrical insulator in which no evidence for the anomaly has been detected to date is amorphous arsenic [3.23b,c]. From an analysis of their data obtained above 0.35 K, these authors put an upper limit on the linear term of $c_1 = 0.5 \pm 0.5$ erg/g K^2. Another peculiarity they noted is that the peak in the C/T^3 vs T plot, which occurs at 5 K in bulk As, is greatly reduced and shifted to 8 K in thin film amorphous As (data in [3.23d]), which demonstrates the sensitivity of the low-temperature specific heat of this substance on its mode of preparation.

The absence of a linear specific heat anomaly was also reported for thin film amorphous germanium [3.24]. On sputtered 20 μm thick films, KING et al. found a specific heat proportional to T^3; the measurements, however, were not extended below 2K (0.5% of θ), and these authors could only put an upper limit of $c_{1,max}$ ≤ 10 erg/gK^2 on the linear term. Amorphous SiO_2 and Se have, in fact, c_1 values very close to that, and hence, the results on amorphous germanium cannot be considered conclusive. c_3, however, was found to be 70% larger than in crystalline Ge; moreover, the hump in the specific heat (ascribed to the higher density of states of the lowest TA phonon branch) occurred at a roughly 20% lower temperature than in the crystalline phase, and was also some 50% larger, i.e., the two phases showed a behavior very similar to that shown in Fig.3.3.

Ewert and co-workers and Buckel and co-workers have searched for a linear specific heat anomaly in thin films of amorphous metals produced by quenched evaporation. Measurements above 0.5 K have yielded no evidence for a linear term. COMBERG et al. [3.25] estimated that the anomaly, if it existed, would have a $c_1 < 80$ erg/g K^2 in superconducting amorphous indium (containing 20 at % antimony). This limit, however, is larger than the largest value of c_1 observed in any of the dielectric glasses. KÄMPF and BUCKEL [3.26] did not estimate an upper limit of c_1 in the amorphous phase of superconducting $Pb_{70} Bi_{30}$, but we estimate that the upper limits on c_1 based on their work is of the same magnitude as determined by EWERT and co-workers. Work below 0.5 K is currently in progress, and the discovery of a linear specific heat term in amorphous bismuth has recently been reported [3.26a].

GRAEBNER et al. [3.27] published the first evidence for a linear specific heat anomaly in an amorphous metal. The superconducting $Zr_{0.7}Pd_{0.3}$ ($T_c = 2.53$ K) was prepared in bulk form as a ribbon of cross section 0.085 × 0.0032 cm. Their data are shown in Fig.3.9. The dashed line, which fits the data around 0.1 K, ($\ll T_c$) corresponds to a linear term with $c_1 = 10.6$ erg/g K^2, very similar to the c_1 terms found in insulating glasses. The lattice T^3 term near and below 1 K is masked by unpaired electrons. $c_3 = 34.9$ erg/g K^4 determined at higher temperatures (~ 10 K) cannot be compared with the Debye prediction, since apparently no speed of sound

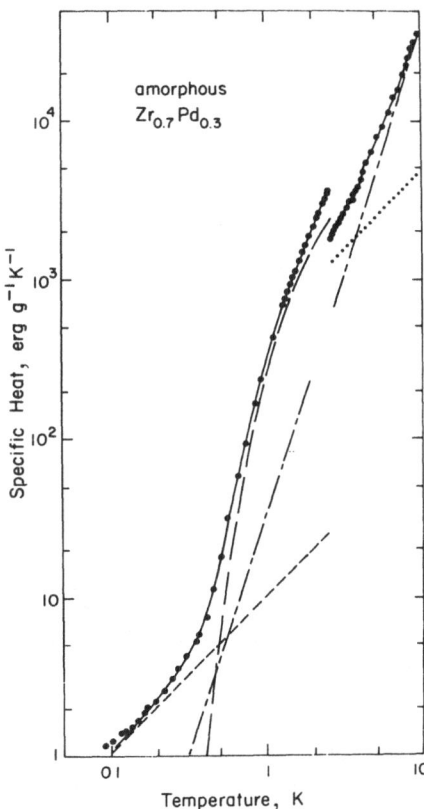

Fig.3.9. Specific heat of amorphous $Zr_{0.7}Pd_{0.3}$ produced as ribbon from the melt [3.27]. The superconducting transition at $T_c = 2.53K$ is clearly shown. The exponential drop of the normal electrons is shown with the long-dash curve. The short-dash line is described by $C = (10.6$ erg gm $K^2)T$. The long-short dashed line is a T^3 extrapolation of the specific heat near 10K. The dotted line represents the electronic specific heat above T_c

measurements exist for this material. It should be noted, however, that GOLDING et al. [3.28] found a c_3 term very close to c_{Deb} in amorphous $Pd_{0.775}Si_{0.165}Cu_{0.06}$ ribbons; evidence for a c_1 term in this glass, however, could not be detected with certainty since this metal is not superconducting. In the same material LASJAUNIAS et al. [3.23a] found a somewhat larger linear specific heat in the amorphous than in the crystalline phase. A clear identification of a glassy anomaly, however, could not be made.

The results reviewed in this section can be summarized as follows: A low-temperature specific heat anomaly appears to be characteristic for all amorphous substances (with the exception of amorphous arsenic); it can be described with the approximate formula (3.1), although in certain materials a different power law definitely prevails at low temperature. In all other materials where this anomaly has not been seen to date, the upper limits put on the linear term (c_1) in these experiments have always been larger than the linear terms actually observed in some glasses.

The important question remains: Is the anomaly an intrinsic property of the amorphous state, or is it extrinsic? If it is extrinsic, it appears to be far more difficult to prevent from occurring than extrinsic specific heat anomalies in crystalline solids. Nevertheless, the possibility that the glassy anomaly is ex-

trinsic cannot be excluded with certainty. This ought to be considered in any interpretation of the low-temperature acoustic or thermal anomalies in amorphous solids.
A systematic study of the potential influence of the chemical and physical state on
the specific heat anomaly in amorphous solids is definitely overdue.

A summary of the data given in Table 3.1 is after STEPHENS [3.17], except for
the more recent data on $Zr_{0.7}Pd_{0.3}$ [3.27]. STEPHENS chose, perhaps somewhat arbitrarily in the light of the foregoing review, always the lowest specific heat data
obtained on a given material, and also approximated them with (3.1). Nevertheless,
the listing may be useful for the purpose of discussion. In column 7 and 8 the constants c_1 and c_3 are listed, determined from specific heat measurements. Column 6
shows the low-temperature T^3 terms expected on the basis of the Debye model:

$$C_{Deb} = c_{Deb} T^3 , \tag{3.2}$$

where

$$c_{Deb} = 234 \frac{Nk_B}{\rho \theta^3} = \frac{2\pi^2}{5} \frac{k_B^4}{\hbar^3 \rho v_{Deb}^3} \tag{3.3}$$

and

$$\frac{1}{v_{Deb}^3} = \frac{1}{v_\ell^3} + 2 \frac{1}{v_t^3} ; \tag{3.4}$$

here, N is the number density of atoms, and ρ the mass density of the glass (column
1); v_ℓ and v_t are the experimentally determined longitudinal and transverse sound
velocities, respectively; k_B is Boltzmann's constant; and $\hbar$ is the Planck's constant $h/2\pi$. v_{Deb} is listed in column 4.

The experimentally determined c_3 term is larger than expected on the basis of
the Debye model, as a comparison of columns 6 and 8 shows. The Debye model predicts
a phonon density of states

$$\frac{\partial n(\hbar\omega)}{\partial(\hbar\omega)} = \frac{3(\hbar\omega)^2}{2\pi^2\hbar^3 v_{Deb}^3} = \frac{15\rho c_{Deb}(\hbar\omega)^2}{4\pi^4 k_B^4} = b_{Deb}(\hbar\omega)^2 . \tag{3.5}$$

The observed specific heat indicates a larger density of states

$$\frac{\partial n(\hbar\omega)}{\partial(\hbar\omega)} = a + (b + b_{Deb})(\hbar\omega)^2 , \tag{3.6}$$

where we have chosen to keep the Debye phonon density of states separate (the justi-
fication for this rests on the fact that acoustic and light scattering experiments
have given evidence for the existence of well-defined Debye phonons in several
glasses). The connection between a and c_1, and between b and c_3 can be worked out
for specific models. One finds the relations [3.22]

$$a = \alpha\rho c_1/\pi^2 k_B^2 , \tag{3.7}$$

Table 3.1. Summary of data available on low-temperature specific heat and thermal conductivity of noncrystalline solids[a]

Material	1 Mass Density g/cm^3	2 Average Atomic Weight	3 Average Molecular Weight	4 v_{Deb} [10^5cm/s]	5 θ_{Deb} (K)	6 c_{Deb} [erg/g K^4]	7 c_1 [erg/g K^2]	8 c_3 [erg/g K^4]	9 n 10^{32} [erg^{-3}cm^{-3}]	10 b_{1064} [erg^{-3}cm^{-3}]	11 n [10^{16}cm^{-3}]	12 β [10^{-4}W cm^{-1}K^{-1}]	13 δ
SiO_2-Spectrosil	2.2	20	60	4.1	494	8.0	12	18	8.42	0.2666	12.4	2.4	1.87
$3SiO_2 \cdot Na_2O$	2.4	20	61	3.51	436	11.8	21	31	16.1	0.573	23.8	1.7	1.92
Corning code No. 7740[b]	2.2	20	...	3.65	440	11.4	~10[e]	22[e]	7.0	0.282	10.5	2.1	1.92
Corning code No. 9700[c]	2.2	20	...	...	...	...	15.0	36.2	10.53	0.661	16.4	...	...
GeO_2	3.6	35	105	2.6	307	19.3	9	26	10.3	0.292	15.1	3.7	1.91
As_2S_3	3.2	49	246	1.69	171	79.0	4.4	97.7	4.5	0.725	8.26	17	1.92
B_2O_3	1.8	14	70	2.04	259	79.9	7.8	90.4	4.5	0.229	6.81	3.5	1.96
Se	4.3	79	79	1.19	113	170	6.6	199	9.05	1.51	16.8	7.9	1.81
Polymethyl-methacrylate	1.2	6.5	104	1.79	256	177	48	292	18.3	1.67	30.0	3.3	1.81
Polystyrene	1.0	6.7	100	1.67	223	262	53	457	16.8	2.36	30.0	2.0	1.87
Lexan[f]	1.2	7.4	238	1.53	210	284	38	410	14.5	2.12	26.1	~2	...[f]
$CaK(NO_3)_3$	2.1	20	217	1.94	230	79.6	39.8	148	26.7	1.74	41.8	1.6	1.90
GE No. 7031[d] varnish	...	...	...	...	...	...	64.5	193	...	...	...	...	...
Pyroceram	2.42	19.8	63.6	4.12	614	7.2	16.0	26.19	27.0	...[k]	...[k]	...	...
$Zr_{0.7}Pd_{0.3}$	7.86[h]	...	...	...[i]	...	...	10.6	34.9	27.0	...[k]	...[k]	8.6	1.9

[a]Largely taken from STEPHENS [3.17]. If one wants more information about the samples not produced for that work, or the source of the numbers in columns 1 to 6, see [3.22]. For Zr Pd, see [3.22]. [b]Nominal composition: 80.5% SiO_2, 38% Na_2O, 12.9% B_2O_3, 2.2% Al_2O_3, 0.4% K_2O, 0.2% Li_2O. 100 ppm Fe estimated from specific heat [3.30]. [c]Nominal composition: 80% SiO_2, 13% B_2O_3, 5% Na_2O, 2% Al_2O_3; 12 ppm Fe estimated from specific heat [3.30]. [d]See [3.29]. [e]From measurements in a 90 kG field and for $T>1.2$K; see [3.19]. [f]After CIELOSZYK et al. [3.31]. The measurements were made down to 0.4 K; this was not low enough to determine the low temperature limit of the temperature dependence of the conductivity. [g]A large T^2 term, $c_2 = 111$ erg/g K^3 was also necessary for a good fit. [h]2% lower than in recrystallized form. [i]No speed of sound data available. [k]Not available since b (column 10) is not known (because v_{Deb} is not known).

$$b = \frac{\gamma \rho}{\pi^4 k_B^4} (c_3 - c_{Deb}) \quad , \tag{3.8}$$

where the constants α and γ depend only slightly on the model. For two-level systems ("tunneling states"), $\alpha = 6$ and $\gamma = 30/7 \approx 4.3$; for harmonic oscillators (infinite-level systems), $\alpha = 3$ and $\gamma = 15/4 \approx 3.8$, and for intermediate number of levels, α and γ have intermediate values. Columns 9 and 10 in Table 3.1 list the values of a and b calculated for two-level states. n, in column 11, is the number density of these states, integrated over a certain energy interval (0.5 to 1.5 k_BK); they are the states which would be involved in the phonon scattering at 0.3K.

Columns 12 and 13 refer to the thermal conductivity which is discussed in a different chapter. It should be kept in mind that almost none of the sample dependence observed in specific heat measurements has been seen in the thermal conductivity, which therefore appears to be closer to a truly intrinsic property. At temperatures below 1 K, the thermal conductivity of all glasses is described by $\varkappa = \beta(T/K)^\delta$. β and δ are listed in columns 12 and 13. Note in particular that δ is not two, although it apparently is very close to it [3.31a].

3.2 Comparison with Theoretical Models

A number of theoretical models have been proposed to explain the specific heat anomaly in amorphous solids. In this section, we will restrict ourselves to a review of several experiments which have been performed in order to test specific predictions made by certain models, i.e., the tunneling model proposed by PHILLIPS [3.32] and independently by ANDERSON et al. [3.33], and the cellular model proposed by BALTES [3.34].

3.2.1 The Tunneling Model

Tunneling is one mechanism by which low excitation energies can be generated. Tunneling of atoms or molecules in crystalline solids is a well-known, although not very common, phenomenon [3.35]. In the tunneling model proposed for glasses [3.32, 33] it has been assumed that a spread of asymmetries and barrier heights between two neighboring potential wells, as might be expected in amorphous materials, could give rise to an essentially constant density of states as required to explain the observed low-temperature linear specific heat anomaly. These tunneling states are also believed to be responsible for the low-temperature thermal conductivity and ultrasonic attenuation. In passing, it ought to be emphasized, however, that no plausible argument has been presented yet why all amorphous substances have approximately the same density of states of tunneling defects which, and that is probably even more puzzling, scatter the phonons with almost equal strength (see Table 3.1). In our opinion, this presents the major theoretical challenge.

The first experiment we will review deals with one aspect of the tunneling states in a rather qualitative way. As was shown in Fig.3.3, the specific heat of solids is generally larger in the amorphous than in the crystalline phase. The hump in the C/T^3 plot occurs at a lower temperature. This has been explained through a softening of the transverse acoustic phonon branch near the Brillouin zone boundary. According to the tunneling model, the linear specific heat anomaly at lower temperatures in the amorphous phase can be similarly viewed as arising from a small fraction of very weakly bound, or tunneling, atoms or molecules. It would appear likely that, if the amorphous solid could somehow be "hardened", both the amorphous hump and the linear anomaly of the specific heat would vanish. In Fig.3.10 (taken from [3.36]), the upper and the lower solid curves represent the specific heat of silica and α-quartz, respectively, plotted as C/T^3. Note the shift towards lower temperatures, and the increase of the hump in the amorphous phase. Irradiation of both phases with fast neutrons [3.36-39] appears to produce states which are intermediate between the crystalline and the glassy states, as evidenced by the shift of the hump, and its changing magnitude. The irradiation is also known to change the mass density [3.40]. The unirradiated silica has a density 85% that of α-quartz. Extended neutron irradiation of the glass produces a density approximately 88% that of quartz (while it lowers the mass density of the crystal). Hence, both specific heat and mass density indicate the creation of a state intermediate between the amorphous and the crystalline states. It would stand to reason that this would also decrease the density of weakly bound (tunneling) states and, hence, decrease the linear specific heat anomaly. The change of the specific heat near 1 K, however, is only slight; in Fig.3.10, there actually appears to be an increase. However, below 1 K, SMITH et al. [3.40a] and RAYCHAUDHURI et al. actually observed a decrease of as much as ∼ 20% together with a decrease of the thermal resistivity by the same percentage. These observations indicate that in the more compact irradiated silica, fewer anomalous (or tunneling) states occur, and that the anomalous states seen in both measurements are closely related.

The second study we will review involves a recent theoretical investigation of the internal consistency of the tunneling model and some very recent work on specific heat measurements on the time scales of microseconds.

The phenomenological tunneling model has been used to successfully describe a wide variety of thermal, ultrasonic and dielectric observations. Recently, BLACK and HALPERIN have performed a very detailed and quantitative comparison of the predicted and observed low-temperature specific heat anomaly [3.41-43]. In the following we will briefly summarize some of their most important findings.

Instead of postulating, as had been done in earlier work, that the density of the tunneling states involved in the interactions with phonons is given by the

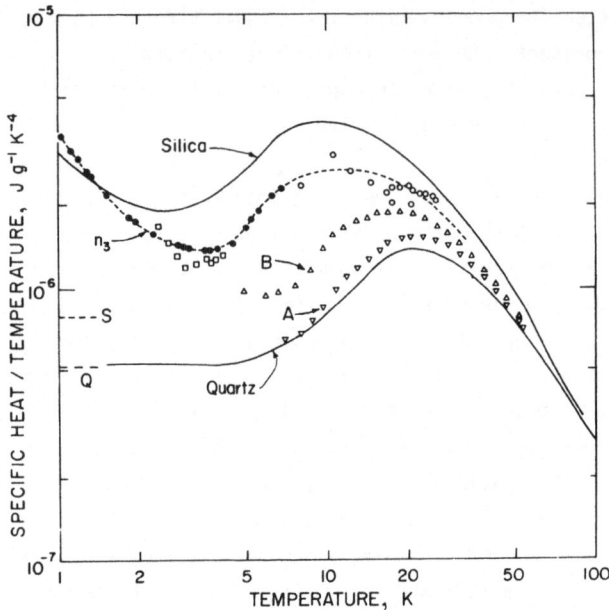

<u>Fig.3.10.</u> Specific heat of silicon dioxide above 1K, plotted as C/T^3. Solid curves, unirradiated silica and quartz, respectively (after [3.9]). Curves A and B, fast-neutron irradiated quartz: $A-2.5 \times 10^{19}$ cm^{-2}; $B-7.7 \times 10^{19}$ cm^{-2} (after [3.37]). Curve n_3, irradiated silica, closed circles [3.36], and open squares [3.38], same sample. Open circles: Different sample, but same fluence (5×10^{19} cm^{-2}) (after [3.39]). Dashed lines labelled S and Q are Debye specific heats calculated from measured sound velocities of silica and quartz, respectively

specific heat [see (3.7)], BLACK and HALPERIN attempted to predict this density of states from ultrasonic observations and to compare it with the measured specific heat density of states. In the tunneling model, the phonon mean free path ℓ resulting from resonant absorption and emission is given by

$$\ell_\alpha^{-1} = (\pi\omega/\rho v_\alpha^3)\bar{P}\gamma_\alpha^2 \tanh(\hbar\omega/2k_BT) \quad , \tag{3.9}$$

where $\alpha = \ell$, t designates longitudinal or transverse phonon polarization, γ_α is the deformation-potential tensor and $\bar{P}$ is the density of states of the most strongly coupled tunneling states. According to the theory, the connection between $\bar{P}$ determined from ultrasonic attenuation and $dn(\hbar\omega)/d(\hbar\omega)$ observed in specific heat measurements is given by

$$dn(\hbar\omega)/d(\hbar\omega) = \bar{P} \min[\eta, \ln 4R_{max} \, t] \quad ; \tag{3.10}$$

here, η is defined as

$$\eta = \lambda_{max} - \lambda_{min} + \ln 2 \quad , \tag{3.11}$$

where λ is a parameter describing the wave function overlap between the states in the two neighboring wells; λ_{max} characterizes maximum isolation in a well.

R_{max}, the maximum (tunneling state) — (phonon bath) relaxation rate for the thermally excited states (with energy of the order $\hbar\omega \approx 2k_BT$), is obtaind from the expression of the (tunneling state) — (phonon bath) relaxation rate

$$R(\hbar\omega,\lambda) = \left(\frac{\gamma_\ell^2}{v_\ell^5} + 2\,\frac{\gamma_t^2}{v_t^2}\right)\frac{(\hbar\omega)^3\ e^{-2}(\lambda\ -\ \lambda_{min})}{2\pi\hbar^4\rho}\ \coth\!\left(\frac{\hbar\omega}{2k_BT}\right)\ , \qquad (3.12)$$

by setting $\lambda = \lambda_{min}$ (i.e., symmetric wells, for which the level splitting $\hbar\omega$ is entirely the result of the tunneling).

$\bar{P}\,\eta$ is the total density of tunneling states observed in specific heat measurements performed on time scales long enough that even the slowest states can equilibrate, i.e., $\eta < \ln 4\ R_{max}t$. If, however, the experiment is performed on a time scale of the order t, such that $\ln 4\ R_{max}t < \eta$, the specific heat measurement will only observe $\bar{P}\ (\tfrac{1}{2}\ln4\ R_{max}t)$ states.

From ultrasonic measurements BLACK and HALPERIN argued that $\bar{P}$ and γ_α are both separately known for silica, and, hence, $dn(\hbar\omega)/d(\hbar\omega)$ can be predicted from (3.10). The density of states determined in this way, however, was found to be 2-3 times smaller than that obtained in specific heat measurements performed on long-time scale (transient heat pulses, of the order of 10 s sample-to-bath thermal time constants) [3.42].

BLACK and HALPERIN also noted that if the specific heat is measured on shorter time scales, the discrepancy becomes even worse. As can be seen from (3.10), a heat pulse diffusing through a thin glass sample (i.e., in a short time t) should encounter only a smaller density of tunneling states. Such an experiment has been performed by GOUBAU and TAIT [3.44] on the time scale of 10^{-4} s. Instead of a considerable overshoot during the first 10 µs, as expected on the basis of the model, and a temperature rise corresponding to a specific heat 7 times smaller than predicted on the basis of ultrasonic data for times of several hundred microseconds, GOUBAU and TAIT observed heat pulses (at T > 160 mK) roughly as expected on the basis of both the experimental long-time specific heat and the thermal conductivity. By varying $\bar{P}$ and γ_α by as much as (and even more than) compatible with the uncertainty in the ultrasonic data, without reaching satisfactory agreement with the experiment, BLACK and HALPERIN finally concluded that a single set of "standard" tunneling states was incapable of describing the short time specific heat results. They explored the possible existence of additional tunneling states, "anomalous" ones in the sense that none of them would couple to the phonons strongly enough to contribute noticeably to their scattering, while all of them should couple sufficiently strongly to contribute to the specific heat on the time scale of Goubau and Tait's experiment. This, however, they found to require a relaxation time for the anomalous states of the order of ~10 µs at 160 mK. Such a short relaxation time, in turn, requires a large γ, and, hence, could lead to noticeable effects in thermal conductivity. We will return to this point.

Although the uncertainties in the experimental data and the number of adjustable parameters are large enough to avoid obvious inconsistencies, it was quite clear from Black and Halperin's work that specific heat work on even shorter time scales would provide a crucial test for the tunneling model in the original as well as in its expanded form including "anomalous" states. Such work has recently been done.

Using heating by very short light pulses and employing faster temperature recording technique, KUMMER et al. [3.45] were able to observe heat pulses in silica at temperatures as low as 50 mK and at times less than 1 μs. Figure 3.11 shows the data taken from their report. Just like those obtained by GOUBAU and TAIT, their data can be described quite satisfactorily with the diffusion of heat, using the specific heat and thermal conductivity measured on long time scales (the deviation from the diffusion profile observed in Fig.3.11c,d can be explained through the transparency induced by saturation of the tunneling states).

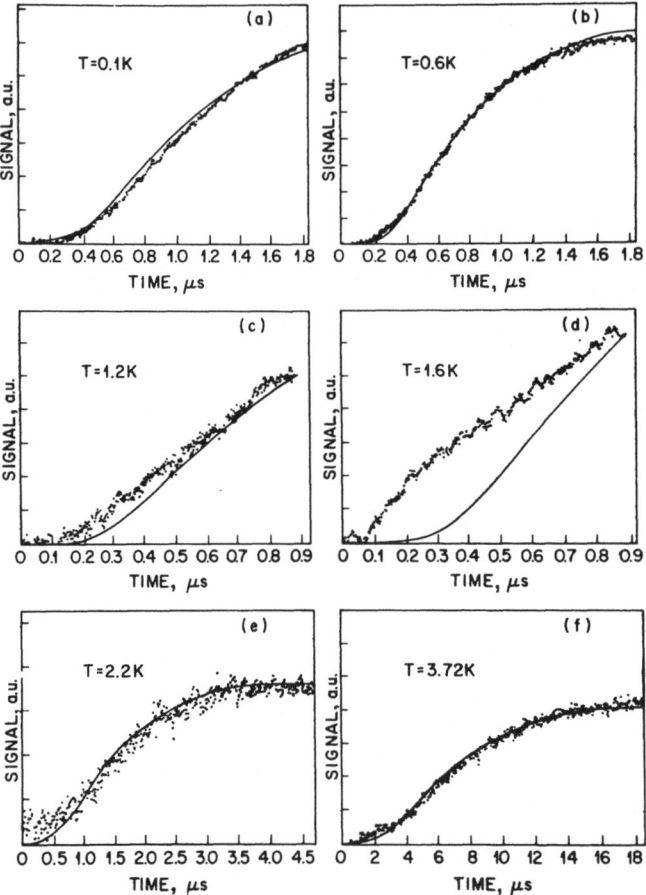

Fig.3.11. Arriving heat pulse shape as a function of time for a 0.127 mm thick silica sample [3.45]. The solid curves are fits to the data, using the standard one-dimensional diffusion equation, with the diffusivity adjusted for best fit

Using the theory developed by BLACK and HALPERIN, RAYCHAUDHURI [3.46] has calculated the phonon mean free path that is to be expected for phonon scattering by the "anomalous tunneling states" which couple to the phonon bath with a 1 μs relaxation time, which would be the upper limit for the relaxation times required to explain the heat pulse data by KUMMER et al. at 0.1 K. Following the analysis given by BLACK ([3.41], see in particular Sect.2.C.1) RAYCHAUDHURI determined that the anomalous levels alone would cause a phonon mean free path $\ell_t \sim 2 \times 10^{-2}$ cm for the transverse dominant phonons at this temperature, and $\ell_\ell \sim 3 \times 10^{-3}$ cm for the longitudinal ones. These values are comparable to the phonon mean free path $\bar{\ell}$ determined from the thermal conductivity using the gas kinetic expression for the thermal conductivity $\varkappa = 1/3\ C_{Deb}v_{Deb}\bar{\ell}$, which is $\bar{\ell} = 2 \times 10^{-2}$ cm [3.9]. In other words, in order for the anomalous states to respond as rapidly as required to explain the heat pulse data, they would have to have a major influence on the thermal conductivity. This is inconsistent with the model.

In an extension of the short-time specific heat measurements by KUMMER et al., LOPONEN et al. [3.45a] have observed that on time-scales of 1-10 ms the low-temperature specific heat of amorphous SiO_2 is smaller than on time-scales exceeding 100 ms. The long-term (> 100 ms) specific heat was found to agree with that reported for silica from "d.c." experiments, and displayed the linear specific heat anomaly. The smaller specific heat observed on the short-time scales varied as T^3; it was, however, larger than the Debye T^3 term. It is not clear how the relatively narrow range of relaxation times observed for the states involved in the linear specific heat anomaly can be reconciled with the picture that these states result from a tunneling process. The suggestion by LOPONEN et al. that it may be the states associated with the anomalous T^3 specific heat that are responsible for the phonon scattering, however, is even more provocative and requires further studies.

A time-dependent specific heat has also been reported by LEWIS et al. [3.45b]. Using a diffusion technique, these authors reported evidence for a time-dependent specific heat on time scales of 100 ms (at ~ 100 mK). RAMMAL and MAYNARD [3.45c] interpreted these results as indications for two kinds of tunneling states, fast relaxing ones and slow ones, the former being 28 times more numerous. The slow ones are the ones which were found to behave in agreement with the tunneling model.

The present experimental situation of short-term specific heat measurements is too confused to yield a coherent picture. Efforts at a critical evaluation are underway [3.45d].

3.2.2 The Cellular Model

This model, proposed by BALTES [3.34], is based on the assumption that amorphous substances may be sufficiently inhomogeneous that their elastic excitations can be approximated as those of macroscopic grains held together by weak bonding forces.

BALTES and HILF [3.47] had calculated the specific heat enhancement ΔC of such an
assembly over the specific heat predicted by the Debye model based on an average
sound velocity. It consisted of two terms:

$$\Delta C = \frac{1}{6} \frac{k_B^2}{\hbar v} \frac{T}{R^2} + \frac{9\zeta(3)k_B^3}{4\pi\hbar^2 v^2} \frac{T^2}{R} \qquad (3.13)$$

where v is the average speed of sound, R the radius of the particle assumed to be
spherical, and ζ the Riemann ζ-function. The first term is the important one, since
it varies as T; its physical origin is the curvature of the particle surface. The
second term results from the surface itself. Although (3.13) was calculated for
spherical particles, calculations for other shapes would change the expression
only by factors of order unity [3.34].

In applying this model to the measured specific heat anomaly in silica and ger-
mania, BALTES found that he had to assume average particle radii $R \approx 30$ Å, and
pointed to dark-field electron microscopic evidence for such a granularity in
glasses. One important inconsistency in Baltes' treatment resulting from such small
cell sizes has been pointed out by TAIT [3.48]. He observed that in materials like
SiO_2 the dominant phonon wavelength at 0.1 K is of the order of 10^4Å, and hence the
excitations involving the particle curvatures at these temperatures would have to
extend over hundreds of cells. It is not clear whether such extended excitations
can still be meaningfully described by the cellular model.

Nevertheless, the question how granularity affects the low-temperature specific
heat is a very interesting one and was studied by TAIT [3.48] on a number of materials. Figure 3.12 shows the results obtained on small particles of (crystalline)
Al_2O_3. Pellets pressed of 200 Å diameter particles to a 50% filling factor showed
a distinct increase of the specific heat over that of single crystal Al_2O_3 (measured
in [3.49]). The same enhancement was observed on two samples prepared from particles obtained from two different manufacturers; this suggests that the specific heat
enhancement is not caused by accidental chemical impurities in the powders. This
conclusion is further supported by the observation that after sintering at 1500°C,
the specific heat of the Linde B material was found to be very close to that of
single crystal Al_2O_3 (Fig.3.12). In the sintered sample, the average particle size
had grown to ~ 0.8 μ, and the filling factor had increased to 95.8%. Subsequent
annealing (at ~ 1800°C, d = 20 μ, 98,5% filling factor) was found to have no influence on the specific heat. TAIT concluded that the residual anomaly found in
the sintered samples below 2 K could be accounted for with 7 ppm of magnetic impurities, on the basis of the total entropy of the Schottky anomaly, assuming two-
level systems. According to the manufacturer, Linde B contains ~15 ppm iron.

The anomaly in the pressed samples (prior to sintering) varies linearly with
T at low temperatures, and its magnitude is close to the anomaly characteristic
for bulk glasses. The anomaly extends to the highest temperatures measured (20 K),

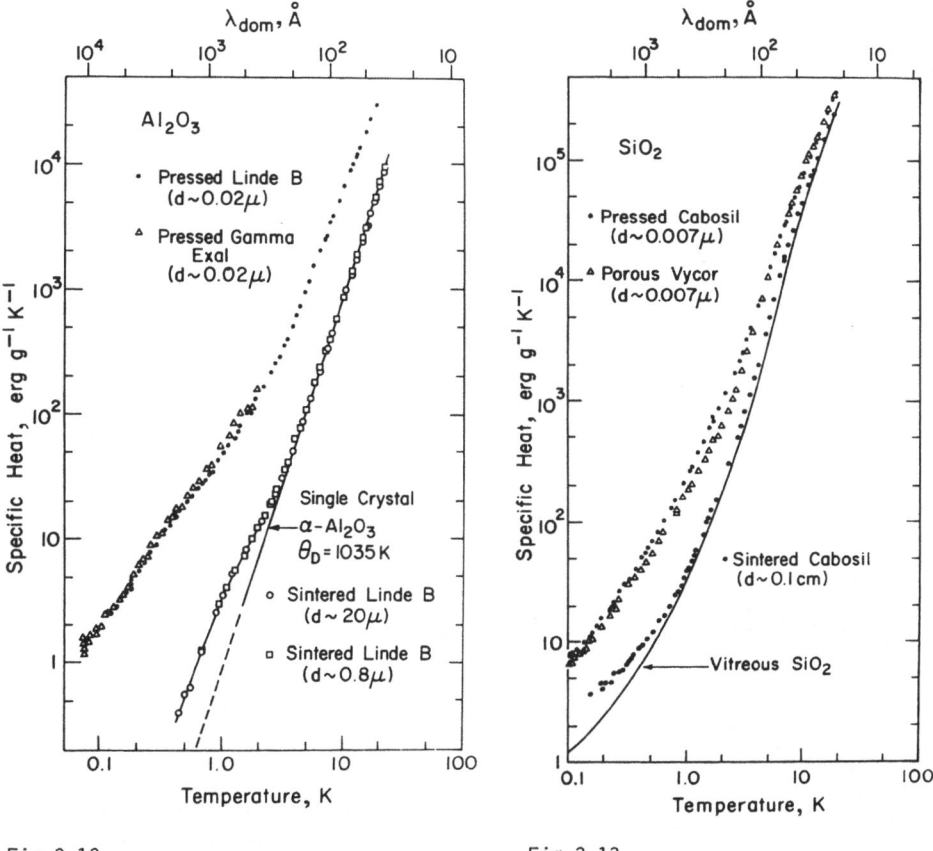

Fig.3.12   Fig.3.13

Fig.3.12. Specific heat of two samples of pressed Al₂O₃ powder obtained from two
different sources. Sintering the Linde B samples produces a specific heat close to
that of single crystal Al₂O₃ [3.49] (after [3.48])

Fig.3.13. Specific heat of pressed silica powder (Cabosil) and of porous Vycor
glass (after [3.48])

where it exceeds the specific heat of single crystal Al₂O₃ by approximately
fivefold.

Similar results were found in pressed Cabosil, which consists of amorphous
silica particles of 70 Å average diamter (53% filling factor), as shown in Fig.3.13.
Surprisingly, this specific heat, which exceeds that of bulk vitreous silica, was
found to be equal to that of porous Vycor, which can be viewed as an interconnected network of silica with pore sizes of ~70 Å, and an average density 70% of
bulk silica; it is produced by leaching the boron-rich phase out of a phase-sep-
arated borosilicate glass in hydrochloric acid. Subsequent sintering of both the
pressed Cabosil and the Vycor resulted in bulk silica with its well-known specific
heat (Fig.3.4).

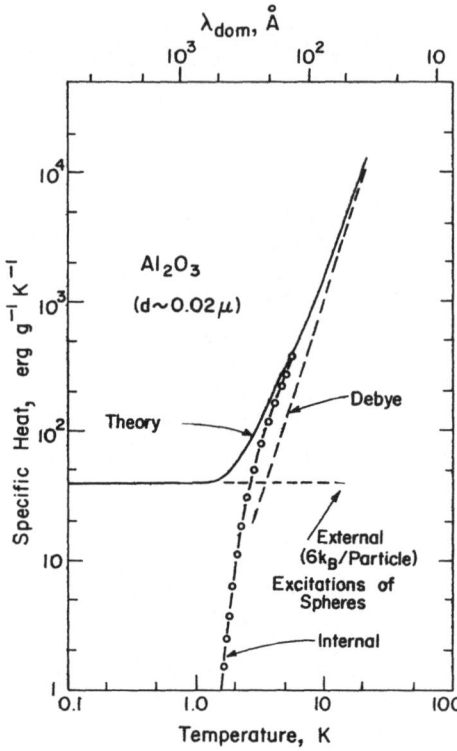

Fig.3.14. Experimental and theoretical specific heat of packed Al_2O_3 (d~0.02) powder (after [3.48])

TAIT attempted to describe this specific heat enhancement with the theory developed by BALTES, although he did use a cutoff for excitations with wavelengths exceeding the particle sizes, resulting in the circles in Fig.3.14. Note that they fall short of the experimental results over the entire temperature range.

An additional mechanism for the excess specific heat of sufficiently loosely bound particles is a vibrational oscillation of the rigid particle [3.50]. If $\hbar\omega_{particle} \ll k_B T$, the specific heat would be in the Dulong-Petit limit 6 $nk_B T$, were n is the number of particles divided by the mass of the pressed sample. This contribution to the specific heat is also indicated in Fig.3.14. Through the proper choice of the distribution of force constants, the observed low-temperature linear specific heat anomaly could probably be described reasonably well. TAIT argued, however, that it would be difficult to explain why the density of states would be the same in Cabosil and in Vycor, since the latter substance is formed by interconnected silica fibers which are likely to be bound far more tightly than the particles in the pressed powder. For this reason, one would expect the force constants and the density of states $dn(\hbar\omega_{particle})/d(\hbar\omega_{particle})$ to be quite different in the two cases. Consequently, TAIT concluded, this mechanism als offers no satisfactory explanation for the anomalous specific heat.

Besides surface phonons, rigid particle oscillations and bulk impurities, TAIT also considered the possibility of weakly bound atoms or molecules on the surface

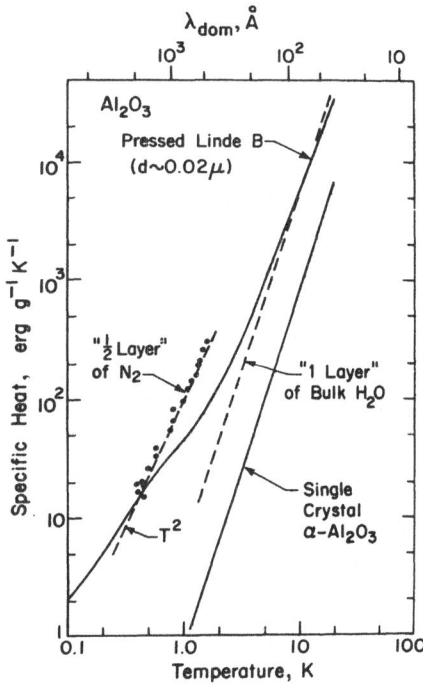

Fig.3.15. Specific heat of monolayer of adsorbed N_2 measured on pressed Alon (.02μ) compared to the specific heat of the sample of Linde B (.02μ). The specific heat of the H_2O adsorbed on the surface of the particles is calculated under the assumption that the specific heat of a monolayer of water adsorbed on the pressed Al_2O_3 sample equals that of the same amount of water in bulk form [3.48]

of the grains; conceivably, adsorbed impurities could also be responsible. The surface area of the porous granular materials of 200 Å particle diameter is ~ 100 m^2/g, 70 Å particles have ~400 m^2/g, and porous Vycor has a measured surface of ~130 m^2/g. TAIT estimated that a sample with 100 m^2/g could accommodate approximately 7×10^{20} molecules of H_2O or N_2 in a completely filled monomolecular layer. Linde B (d = 0.02 μ) which had been exposed to air, for example, has been shown to pick up that much water [3.51], which could not be completely removed through outgassing even at high temperatures (1000°C).

The influence of such adsorbed molecules on the specific heat was demonstrated by TAIT. During cool-down he absorbed onto a pressed sample of d = 200 Å alumina particles what amounted to an estimated half monolayer of N_2. The specific heat enhancement, labelled "½ layer" of N_2, shown in Fig.3.15, is seen to vary as T^2 as expected for a two-dimensional solid. At 1 K, the measured specific heat is approximately equal to that of an equal amount of nitrogen in bulk form. For the purpose of comparison, Fig.3.15 also shows the T^3 specific heat of bulk H_2O corresponding to one monolayer (7×10^{20} molecules per gram Al_2O_3) on 200 Å diameter alumina powder. Although Fig.3.15 by no means should be considered as proof that the excess specific heat of the fine powder observed by TAIT (Figs.3.12,13) is caused by adsorbed molecules, it certainly demonstrates the potential severe limitations of specific heat measurements on samples with large surfaces.

An approach to testing the cellular model without encountering the surface problem is to study the specific heat changes occurring as a result of phase separation or of crystallization in bulk glasses. In particular, the devitrification should provide insight into the changes occurring as the structure changes from the amorphous to the crystalline state containing increasingly larger crystallites. Although much work has been done on this subject, the experimental situation is still not clear. We will review this work in the following.

Heat treatment of silica at 1400-1500°C causes it to devitrify; the resulting crystal phase is that of (polycrystalline) cristobalite. The crystallites have diameters of a few microns, and the material is described as highly porous [Ref. 3.52, p.60]. Figure 3.16 shows the specific heat of cristobalite as measured by BILIR and PHILLIPS [3.53] between 2 and 22 K. Apparently, the crystallization removes the low-temperature anomaly; to be precise: the experimental data allow placing an upper limit of $c_1 \leq 10$ erg/g K^2 on the linear anomaly. The hump at $T/\theta \sim 3 \times 10^{-2}$ has been correlated with the soft transverse acoustic mode in cristobalite.

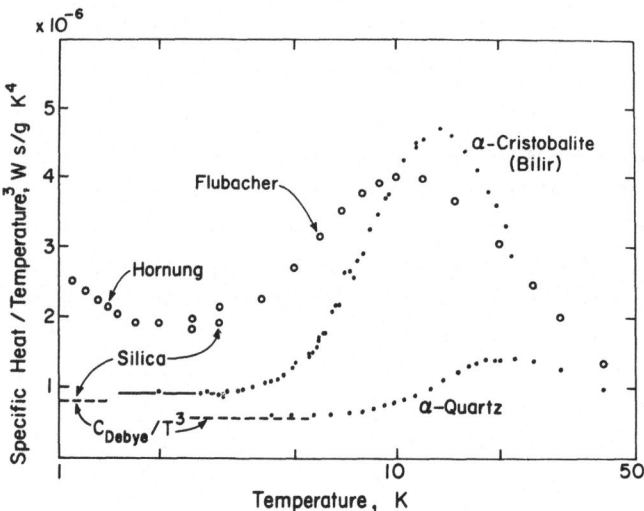

Fig.3.16. Specific heat C/T^3 of silica and α-quartz (same data as shown in Fig.3.4), and of cristobalite [3.52,53]. C_{Deb}/T^3 is calculated from measured speeds of sound of silica and α-quartz, respectively

TAIT searched for changes of the specific heat in the glass ceramic (MgO, Li_2O)-Al_2O_3-SiO_2 (Corning "Pyroceram"). As received, the average crystallite size was 600 Å. The specific heat in that state, measured by STEPHENS and TAIT, is shown in [Ref.3.17, Fig.14]. Below 2 K, the specific heat was masked by magnetic impurities, as shown through changes produced in a 33 kG magnetic field. Above 2 K, the specific heat very closely resembled that of vitreous silica. It exceeds the Debye prediction, based on speed of sound measurement in the Pyroceram, by a factor between 3 and 5 over the entire range of measurement (up to 20 K). Through successive annealings TAIT made the crystallites grow by over a factor of 15, to 1μ diameter. Yet the specific heat remained entirely unaltered over the entire range of measurement, 0.2 to 20 K. The fact that the magnetic impurity dominated specific heat did not change is probably not surprising, since the crystal field splitting should not change much with the crystallite size. However, the fact that even above that temperature range the specific heat was unaltered led TAIT to the conclusion that the specific heat is not measurably influenced by any cellular structure in this material.

WYCHERLY [3.54] and BOHN [3.55] measured similar glass ceramics. WYCHERLY observed an anomaly as given by (3.1) in both the amorphous and the crystalline phase. The linear anomaly was unaltered by the crystallization (particle sizes $\geq 1 \mu$), while the specific heat at higher temperatures was found to increase somewhat. Bohn's measurements indicated an increase of the specific heat upon crystallization by almost a factor of two. In no case was a decrease observed, contrary to the expectation based on the cellular model.

The most puzzling results, however, were reported by JEAPES et al. [3.56]. Figure 3.17 shows the specific heat of GeO_2; in the amorphous phase, C/T^3 shows the hump characteristic for the amorphous phase. The increase at the lowest temperatures signals the linear anomaly. In the polycrystalline phase, the high-temperature hump has disappeared (or it has moved to much higher temperatures). However, the low-temperature, linear specific heat anomaly, which is now far more clearly visible, has remained unaltered. The same results were obtained on Diopside ($CaO \, MgO - SiO_2$). Here, JEAPES [3.57] measured two samples, one containing 14 ppm, the other > 100 ppm iron. At the lowest temperatures measured, the latter sample had a somewhat larger anomaly (by $\sim 50\%$). JEAPES took this as evidence that iron could not be the cause for the anomaly, at least not for its bulk. The findings on these two materials are the opposite of what BILIR observed in SiO_2 [3.52].

In both the crystalline GeO_2 and the Dioposide, crystallization causes porosity on the scale of microns. TAIT [3.48] has estimated that, if the porosity corresponded to that of 1μ compressed particles, adsorbed helium could produce an anomaly of the size observed (not N_2 or H_2O, though), based on the work by TAIT et al. [3.58]. The presence of helium, however, can be excluded in Jeapes' experiments [3.59].

To summarize: the search for systematic changes of the specific heat upon devitrification so far has been inconclusive. However, there has not yet been any evidence that a variation of the crystallite sizes has an influence on the specific heat. Although this negative result cannot be used to disprove the applicability of the cellular model in amorphous solids, it does, nevertheless, deny it one, potentially important, piece of supporting evidence.

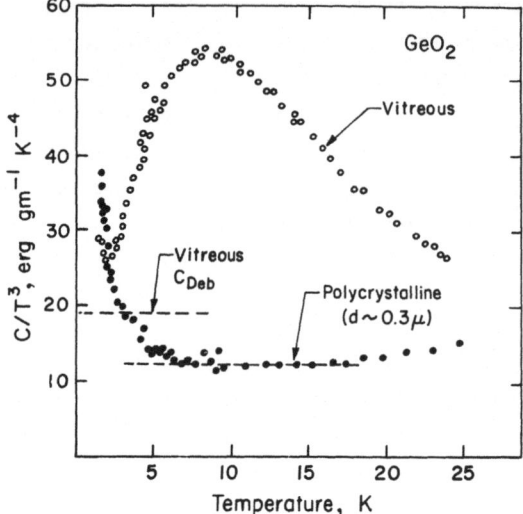

Fig.3.17. Specific heat of vitreous GeO_2 and of recrystallized (tetragonal) GeO_2 plotted as C/T^3 [3.56]. The linear anomaly $c_1 T$ in the amorphous phase has an estimated $c_1 \leq 20$ erg/gmK , in agreement with Stephens' value (Table 3.1): $c_1 = 9$ erg/gmK^2. In the tetragonal phase, the anomaly can be very accurately described with (8 erg/gm K^2)T. A similar anomaly was found in the hexagonal phase of GeO_2. Transition metal impurities can be excluded as cause of anomaly, not only because of the wrong temperature dependence, but also because of the impurity concentrations in the samples, which are too small to account for the total entropy observed

3.3 Summary and Outlook

The specific heat in glasses continues to be a general phenomenon of all amorphous solids—and its origin also continues to be a puzzle. Evidence has been accumulated that the anomaly does not have one single cause; however, a model of two different kinds of excitations, called, for example, the "standard" and the "anomalous" tunneling states by BLACK and HALPERIN, can probably explain the shorttime (heat pulse) specific heat measurements reported in the last few years. The cellular model proposed by BALTES has been tested in a number of experiments on crystallized glasses of different crystallite sizes. Although none of these measurements can be considered conclusive, none of them have given any evidence that a cellular structure in polycrystalline material does give rise to a specific heat as should be expected on the basis of the cellular model.

Much of the work reviewed in this chapter has produced evidence that at least a part of the anomaly is not intrinsic in the amorphous materials. This observation should be taken as a warning, and as an encouragement to the experimenters to pay more attention to the sample preparation. More specific heat measurements on glasses are not likely to shed more light on amorphous solids, unless better characterized samples are employed.

References

3.1   P. Debye: Ann. Physik 39, 788 (1912)
3.2   P.H. Keesom, N. Pearlman: Phys. Rev. 91, 1354 (1953)
3.3   W.T. Berg, J.A. Morrison: Proc. R. Soc. London 242, 467 (1957)
3.4   T.R. Sandin, P.H. Keesom: Phys. Rev. 177, 1370 (1969)
3.5   J.S. Dugdale, J.A. Morrison, D. Patterson: Proc. R. Soc. London A 224, 228 (1954)
3.6   W. de Sorbo, W.W. Tyler: J. Chem. Phys. 21, 1660 (1953)
3.7   W. de Sorbo, G.E. Nichols: J. Phys. Chem. Solids 6, 352 (1958)
3.8   B.J.C. Van der Hoeven, P.H. Keesom: Phys. Rev. 130, 1318 (1963)
3.9   R.C. Zeller, R.O. Pohl: Phys. Rev. B4, 2029 (1971)
3.10 P.H. Keesom, N. Pearlman: Phys. Rev. 91, 1347 (1953)
3.11 R.W. Hill, D.H. Parkinson: Philos. Mag. 43, 309 (1952)
3.12 R.B. Stephens: Ph.D. Thesis, Cornell University (1974); Cornell Materials Science Center Report No. 2304 (1974)
3.13 C.R. Kurkjian: Private communication (Bell Telephone Laboratories, Murray Hill, NJ)
3.14 J.C. Lasjaunias, R. Maynard, D. Thoulouze: Solid State Commun. 10, 215-217 (1972)
3.15 W. Reese, J.E. Tucker: J. Chem. Phys. 46, 1388 (1967)
3.16 A.J. Leadbetter: J. Chem. Phys. 51, 779 (1969)
3.17 R.B. Stephens: Phys. Rev. B 13, 852 (1976)
3.18 J.C. Lasjaunias, A. Ravex, M. Vandorpe, S. Hunklinger: Solid State Commun. 17, 1045 (1975)
3.18a H.V. Löhneysen, B. Picot: J. Phys. Paris 39, Supplement, C6-976 (1978)
3.18b J.C. Lasjaunias, R. Maynard, M. Vandorpe: J. Phys. Paris 39, Supplement, C6-973 (1978)
3.19 R.A. Fisher, G.E. Brodale, E.W. Hornung, W.F. Giauque: Rev. Sci. Instrum. 39, 108 (1968)
3.20 L.E. Wenger, K. Amaya, C.A. Kukkonen: Phys. Rev. B14, 1327 (1976)
3.21 J.C. Lasjaunias, D. Thoulouze: Presented at EPS Low Temperature Conference, Freudenstadt, 1972
3.22 R.B. Stephens: Phys. Rev. B8, 2896 (1973)
3.23 J.C. Lasjaunias, D. Thoulouze, F. Pernod: Solid State Commun. 14, 957 (1974)
3.23a S. Kelham, H.M. Rosenberg: J. Phys. Paris 39, Supplement C6-982 (1978)
3.23b D.P. Jones, N. Thomas, W.A. Phillips: Philos. Mag. B38, 271 (1978)
3.23c D.P. Jones, N. Thomas, W.A. Phillips: J. Phys. Paris 39, Supplement C6-978 (1978)
3.23d C.T. Wu, H.L. Luo: J. Non Cryst. Solids 13, 437 (1974)
3.24 C.N. King, W.A. Phillips, J.P. deNeufville: Phys. Rev. Lett. 32, 538 (1974)
3.25 A. Comberg, S. Ewert, H. Wühl: Z. Phys. B25, 173 (1976)
3.26 G. Kämpf, W. Buckel: Z. Phys. B27, 315 (1977)
3.26a H. Selisky, W. Buckel: Verhandl. DPG (VI) 15, 399 (1980)
3.27 J.E. Graebner, B. Golding, R.J. Schutz, F.S.L. Hsu, H.S. Chen: Phys. Rev. Lett. 39, 1480 (1977)
3.28 B. Golding, B.G. Bagley, F.S.L. Hsu: Phys. Rev. Lett. 29, 68 (1972)
3.28a J.C. Lasjaunias, A. Ravex, D. Thoulouze: J. Phys. F 9, 803 (1979)
3.29 R.B. Stephens: Cryogenics 15, 420 (1975)
3.30 R.C. Zeller, M.S. Thesis: Cornell University (1971); Materials Science Center Report No. 1453
3.31 G.S. Cieloszyk, M.T. Cruz, G.L. Salinger: Cryogenics 13, 718 (1973)
3.31a A.K. Raychaudhuri, J.M. Peech, R.O. Pohl: In *Phonon Scattering in Condensed Matter*, ed. by H.J. Maris (Plenum Press, New York 1978) p.45
3.32 W.A. Phillips: J. Low Temp. Phys. 7, 351 (1972)
3.33 P.W. Anderson, B.I. Halperin, C.M. Varma: Philos. Mag. 25, 1 (1972)
3.34 H.P. Baltes: Sol. State Commun. 13, 225 (1973)
3.35 V. Narayanamurti, R.O. Pohl: Rev. Mod. Phys. 42, 201 (1970)
3.36 A.K. Raychaudhuri: Ph. Thesis, Cornell University (1980); Materials Science Center Report No. 4284

3.37 E.F. Westrum: In *IV Congres International de Verre*, Paris (France) 1959, p.396
3.38 G.K. White, J.A. Birch: Phys. Chem. Glasses *6*, 85 (1965)
3.39 A.E. Clark, R.E. Strakna: Phys. Chem. Solids *3*, 121 (1962)
3.40 W. Primak: *Compacted States of Vitreous Silica*, Studies in Radiation Effects in Solids, Vol.4 (Gordon and Breach, New York 1975) in particular p.85ff.
3.40a T.L. Smith, P.J. Anthony, A.C. Anderson: Phys. Rev. B*17*, 4997 (1978)
3.41 J.L. Black: Ph.D. Thesis, Harvard University (1977)
3.42 J.L. Black, B.J. Halperin: Phys. Rev. B*16*, 2879 (1977)
3.43 J.L. Black: Phys. Rev. B*17*, 2740 (1978)
3.43a J.L. Black: J. Phys. Paris *39*, Supplement C6-963 (1978)
3.44 W.M. Goubau, R.H. Tait: Phys. Rev. Lett. *34*, 1220 (1975)
3.45 R.B. Kummer, R.C. Dynes, V. Narayanamurti: Phys. Rev. Lett.*40* , 1187 (1978)
3.45a M.T. Loponen, R.C. Dynes, V. Narayanamurti, J.P. Garno: Phys. Rev. Lett. *45*, 457 (1980)
3.45b J.E. Lewis, J.C. Lasjaunias, G. Shumacher: J. Phys. Paris *39*, Supplement C6-967 (1978)
3.45c R. Rammal, R. Maynard: J. Phys. Paris, Supplement C6-970 (1978)
3.45d M.T. Loponen, R.C. Dynes, V. Narayanamurti: To be published (footnote in [3.45a])
3.46 A.K. Raychaudhuri: Private Communication (Cornell University)
3.47 H.P. Baltes, E.R. Hilf: Solid State Commun. *12*, 369 (1973)
3.48 R.H. Tait: Ph.D. Thesis, Cornell University (1975); Cornell Materials Science Center Report No. 2454
3.49 R.Q. Fugate, C.A. Swenson: J. Appl. Phys. *40*, 3034 (1969)
3.50 G. Jura, K.S. Pitzer: J. Am. Chem. Soc. *74*, 6030 (1952)
3.51 H. Boutin, H. Prask: Surf. Sci. *2*, 261 (1964)
3.52 N. Bilir: Ph. D. Thesis, Stanford University (1974)
3.53 N. Bilir, W.A. Phillips: Philos. Mag. *32*, 113 (1975)
3.54 K.E. Wycherly: Ph.D. Thesis, University of Bristol (1969)
3.55 R.G. Bohn: J. Appl. Phys. *45*, 2133 (1974)
3.56 A.P. Jeapes, A.J. Leadbetter, C.G. Waterfield, K.E. Wycherley: Philos. Mag. *29*, 803 (1974)
3.57 A.P. Jeapes: Ph.D. Thesis, University of Bristol (1974), private communication by A.J. Leadbetter (data shown in [3.48], Fig.38)
3.58 R.H. Tait, R.O. Pohl, J.D. Reppy: *Low Temperature Physics - LT13*, Vol.1, ed. by K.D. Timmerhaus et al. (Plenum, New York 1975) p.172
3.59 A.J. Leadbetter: Private communication

4. The Thermal Expansion of Glasses

W. A. Phillips

With 3 Figures

Although for practical purposes thermal expansion is one of the more important properties of a glass, there are no satisfactory descriptions of the effect at a microscopic level. The main reason for this is the complexity of the subject. A complete calculation of the thermal expansion requires not only a knowledge of the vibrational density of states, but also an understanding of the way in which each vibrational mode changes with pressure and, to a lesser extent, temperature. Recent work, as described in Chap.2, has shown how the density of states in a real glass can be calculated, but there have been few attempts to calculate pressure effects.

In contrast, measurements of thermal expansion have been made in a large range of glasses over a wide temperature range. This chapter will concentrate on measurements below 4 K, where the results are dramatic and unexpected, but will also describe how some progress has been made in understanding the room temperature behaviour. Almost inevitably the discussion will centre on vitreous silica, for which the most complete data are available, and which conveniently illustrates two different aspects of the physics of thermal expansion.

Firstly, the coefficient of expansion of vitreous silica at room temperature is much smaller than that of the crystalline forms. This 'anomalous' expansion was not understood for many years, but Sect.4.2 describes how the development of a microscopic theory can explain the anomaly. An outline of the theory is given in Sect.4.1. Secondly, at low temperatures the thermal expansion coefficient is negative in SiO_2, and, as described in Sect.4.3.1, below 4 K becomes much larger than can be explained on the basis of normal vibrational modes. This is linked in Sect.4.3.2 to the excess heat capacity (described in Chap.3) using the idea of tunneling states. Other data are also described in Sects.4.2 and 4.3, including measurements in sodium silicate glasses. However, this chapter is not intended as a comprehensive review of thermal expansion in glasses.

4.1 Theoretical Background

In much of the literature measurements of the thermal expansion of solids have been interpreted in terms of the quasi-harmonic approximation. Anharmonicity is taken into account in this approximation by allowing the frequencies ω_i of the normal modes of vibration to vary with the volume V of the solid. No explicit temperature dependence is introduced; ω_i is independent of temperature at fixed V. This approximation is, at best, of doubtful validity. In its simplest form it implies that the temperature dependence of the mode frequencies can be predicted from a knowledge of the bulk expansion coefficient and the volume dependences of the ω_i, a prediction which is not borne out by experiment. However, it appears that this extreme form of the quasi-harmonic approximation is not necessary. Considerable progress can be made in understanding and interpreting thermal expansion data without its use.

BARRON [4.1] has taken anharmonic effects into account by introducing an explicit temperature dependence of ω_i. He showed that to a first approximation the thermodynamic properties of an anharmonic solid may be represented by substituting appropriate temperature dependent frequencies into the quasi-harmonic expressions for the free energy and the entropy. In general the frequencies needed to fit the free energy are not the same as those needed to fit the entropy. However, the frequency distribution that must be used to fit the entropy is the same as that observed in neutron inelastic scattering experiments, which must, of course, also be affected by anharmonicity. This suggests that the entropy should be used as the starting point for the calculation of thermodynamic quantities, as long as it is expressed in terms of an experimental frequency distribution and not one calculated from the harmonic approximation.

The entropy is therefore written in the quasi-harmonic form

$$S = \sum_i s_i[\omega_i(V,T)/T] \qquad (4.1)$$

where s_i is the contribution to the entropy of the mode i, with a frequency ω_i which is a function of both volume and temperature. The summation is over the *measured* frequency spectrum. Differentiation of $s_i[\omega_i(V,T)/T]$ with respect to temperature and volume leads to the two equations

$$\left.\frac{\partial s_i}{\partial V}\right|_T = s_i'(\omega_i/T)\frac{1}{T}\left.\frac{\partial \omega_i}{\partial V}\right|_T \qquad (4.2)$$

and

$$\left.\frac{\partial s_i}{\partial T}\right|_V = s_i'(\omega_i/T)\left(\frac{1}{T}\left.\frac{\partial \omega_i}{\partial T}\right|_V - \frac{\omega_i}{T^2}\right) , \qquad (4.3)$$

where $s_i'(x) = ds_i(x)/dx$. The contribution to the heat capacity from mode i is, from (4.3),

$$C_i(T) = s_i!(\omega_i/T) \left(\frac{\partial \omega_i}{\partial T} \bigg|_V - \frac{\omega_i}{T} \right)$$

which, when combined with (4.1) and (4.2), gives

$$\frac{\partial S}{\partial V}\bigg|_T = \sum_i C_i(T) \frac{1}{T} \frac{\partial \omega_i}{\partial V}\bigg|_T \left(\frac{\partial \omega_i}{\partial T}\bigg|_V - \frac{\omega_i}{T} \right)^{-1}$$

$$= -\sum_i C_i(T) \frac{\partial \ln \omega_i}{\partial V} \left(1 - \frac{\partial \ln \omega_i}{\partial \ln T} \right)^{-1} . \tag{4.4}$$

The factor in brackets represents the first order anharmonic correction to the heat capacity and can, in general, be neglected in the calculation of the thermal expansion. In SiO_2, for example, $\partial \ln \omega_i / \partial \ln T$ is about 10^{-2} [4.2]. Defining the Gruneisen constant for mode i as

$$\gamma_i = - \frac{\partial \ln \omega_i}{\partial \ln V} \tag{4.5}$$

(4.4) becomes

$$\frac{\partial S}{\partial V}\bigg|_T = \frac{1}{V} \sum_i C_i(T) \gamma_i . \tag{4.6}$$

The importance of (4.6) lies in the fact that it allows a calculation of the thermal expansion to be made even though the quasi-harmonic approximation is inadequate. The derivation presented here treats the temperature and volume dependences of ω_i separately, and (4.4) shows that the explicit temperature dependence enters only as a small correction factor to the γ_i [or to the $C_i(T)$].

Experimentally the average Gruneisen parameter γ is evaluated using

$$\gamma = \frac{\beta V}{\chi_T C_V} = \frac{\partial S}{\partial V}\bigg|_T \frac{V}{C_V} , \tag{4.7}$$

where C_V is the heat capacity (of a volume V) evaluated at constant volume, β is the coefficient of cubic expansion and χ_T is the isothermal compressibility. Writing

$$C_V = \sum_i C_i(T)$$

(4.6) and (4.7) give

$$\gamma = \frac{\sum_i \gamma_i C_i(T)}{\sum_i C_i(T)} , \tag{4.8}$$

which allows the experimental value of γ to be computed from the individual mode Gruneisen parameters γ_i.

In the Debye approximation, which treats the modes of vibration as sound waves, the γ_i can be evaluated from a knowledge of the pressure dependences of the elastic constants. This evaluation is not simple for crystals, involving complicated combinations of the third order elastic constants [4.3], but is straightforward for glasses, which can be treated as elastically isotropic solids. Writing the longitudinal and transverse velocities of sound as v_ℓ anc v_t, and the frequency ω_i as $q_i v_\ell$ or $q_i v_t$ where q_i is the wave vector, the Gruneisen parameters are

$$\gamma_i^\ell = - \frac{\partial \ln q_i v_\ell}{\partial \ln V}$$

and

$$\gamma_i^t = - \frac{\partial \ln q_i v_t}{\partial \ln V} \quad .$$

The wave vectors q_i scale as $V^{-1/3}$, so that the γs are independent of the index i, and can be written as

$$\gamma^\ell = \frac{1}{3} + \frac{1}{\chi T} \frac{\partial \ln v_\ell}{\partial P} \tag{4.9}$$

and

$$\gamma^t = \frac{1}{3} + \frac{1}{\chi T} \frac{\partial \ln v_t}{\partial P} \quad ,$$

where P is the pressure. These equations can be rewritten in terms of the pressure derivations of the elastic constants, but since in practice the pressure derivatives of the velocities are accessible experimentally, the form of (4.9) seems to be more convenient. The value of γ can be calculated from (4.8) using the Debye expression for C_i

$$\gamma = (\gamma^\ell/v_\ell^3 + 2\gamma^t/v_t^3)(1/v_\ell^3 + 2/v_t^3)^{-1} \quad . \tag{4.10}$$

4.2 The High Temperature Expansion of Vitreous Silica

A calculation of the thermal expansion of vitreous silica at room temperature and above provides a good example of the use of (4.6). At room temperature the coefficient of cubic expansion β is about 1.4×10^{-6} in vitreous silica, less than that of quartz and other crystalline forms of silica by a factor of about 25. This difference has been used to account for many of the differences shown in other physical properties [4.4]. Since the thermal expansion of quartz is comparable to that of many other solids, considerable efforts have been made to explain the anomalously low expansion coefficient of vitreous silica.

Equation (4.6) shows that the thermal expansion can be calculated from measurements of the variation of the mode frequencies with pressure, using the measured compressibility to calculate the individual γ_i. The summation in (4.6) is over all modes, so that in principle every γ_i must be measured. In the crystal this could be done by measuring the phonon dispersion relations under pressure, but in the glass, where the phonons do not have well defined wave-vector $\underline{k}$, the best available information is contained in the pressure dependences of infra-red and Raman spectra. These spectra, in the absence of $\underline{k}$ conservation, contain contributions from every mode, although the magnitude of the contribution varies from mode to mode.

Although data for the pressure dependence of the infra-red spectrum are available for vitreous silica [4.5,6], it is not easy to extract reliable values of γ. The shift of each peak in the spectrum is not only small (or the order of 0.1% per kbar), but is accompanied by a comparable broadening. Part of this may well arise from non-uniform pressure distributions, but, in addition, there is no reason to suppose that each mode contributing to a given peak has exactly the same γ. This can also give a broadening.

Although different authors disagree on the magnitude of the average γs corresponding to the three main peaks in the infra-red spectrum of vitreous silica they do agree on the sign. The γs for the modes in the peaks at 1100 and 470 cm^{-1} are negative, and positive for the modes at 800 cm^{-1}. These results can be understood on the basis of an analysis by BELL et al. [4.7], who used a physically constructed model to study the atomic motion associated with each mode. Each mode involves both stretching and bending of the silicon-oxygen bond, but, in general, the higher the frequency the greater the contribution made by the bond stretching forces. For example, the highest frequency is found for oxygen motion parallel to a line joining the two silicon atoms. The effect of external pressure is to decrease the bond angle at the oxygen atom, so the contribution of bond stretching to the highest frequency mode is decreased. This frequency, therefore, decreases with increasing pressure, giving a negative γ. Similar arguments can be made for the other modes.

A calculation of γ requires not only the individual γ_i but also the associated heat capacities $C_i(T)$. Infra-red data do not give a reliable estimate for the density of modes associated with each peak in the spectrum, but since the density of states calculation of BELL et al. can be used to give a specific heat in good agreement with experiment (Fig.4.1), these calculated values can be used in the calculation of the thermal expansion. The result of such a calculation is shown in Fig. 4.1. Since the infra-red data do not extend below 400 cm^{-1} the average Gruneisen constants for the two groups of low frequency modes are chosen to obtain agreement at low temperatures, although the high temperature thermal expansion is insensitive to the precise choice.

The agreement between the calculated and the experimental results shows that the small thermal expansion of vitreous silica arises naturally from the alternate positive and negative mode Gruneisen constants, which in turn are explicable in terms of the atomic displacements associated with each mode. The mode Gruneisen parameters of the crystalline forms of SiO_2 are similar to those of vitreous silica and cannot account for the larger expansion coefficients of the crystalline silicas between room temperature and the α-β phase transition. This larger expansion has

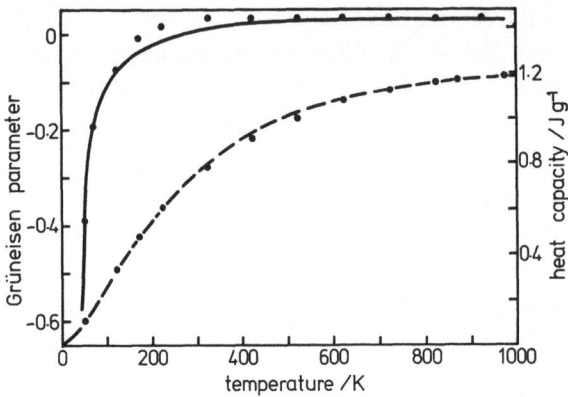

Fig.4.1. Experimental values (solid circles) of the heat capacity and the thermal expansion compared with the corresponding theoretical calculations (dashed and solid lines, respectively)

been shown to be a result of the large volume change at the α-β phase transition, the effects of which extend down to room temperature [4.6].

This use of a direct measurement of γ_i to link the thermal expansion to the modes of vibration of the glass can be extended to more complicated glasses. For example, the addition of Na_2O to SiO_2 increases the thermal expansion [4.8], in contrast to B_2O_3 which has much less effect. The infra-red spectra of the sodium-silicate glasses have been measured under pressure [4.9], and the results show that all the modes have a positive γ_i for the addition of as little as 10% Na_2O. Although no quantitative comparisons have been made, the results are qualitatively consistent and show that in some cases it is possible to use spectroscopic measurements to supplement studies of the effect of glass composition on thermal expansion. Such studies have usually been by empirical albeit elegant methods [4.10].

4.3 The Low Temperature Expansion of Glasses

4.3.1 Experimental Results

The measurement of thermal expansion at liquid helium temperatures is difficult, involving fractional changes in length of less than 10^{-8} K^{-1}. However, experimental difficulties are offset by the hope that (4.10) should apply, so that the experimental Gruneisen constant γ_T can be compared directly to that calculated from the pressure dependence of the sound velocities, γ_S. Of course, this hope relies on the validity of the Debye approximation, and in general is fulfilled in crystalline solids. Just as the heat capacity at low temperatures can be calculated from a knowledge of the sound velocities, so the experimental values γ_T and γ_S agree to within experimental error in the crystal at very low temperatures. This is illustrated in Fig.4.2 where Gruneisen constants derived from experimental values of thermal expansion are plotted against temperature for crystalline germanium [4.11] and potassium chloride [4.12], and compared with the values derived from the

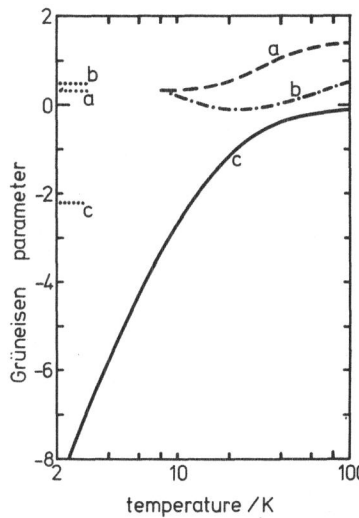

temperature /K

Fig.4.2. The Grüneisen parameter γ_T as a function of temperature for a) potassium chloride,
b) crystalline germanium and c) vitreous silica.
The corresponding calculated values of γ_S are
also plotted as dotted lines

pressure derivatives of the sound velocities [4.12,13] according to (4.9,10). The
agreement is good in the low temperature limit where the Debye description is
adequate, but as expected γ_T departs significantly from γ_S at higher temperatures.

In most glasses the situation is not this simple as γ_T and γ_S are usually
markedly different. This, of course, is not unexpected. Measurements of the heat
capacity show quite clearly that the Debye description is not a good one, at least
down to the lowest temperature for which results are available [4.14]. Below 1 K
the heat capacity is dominated by a large density of additional states, which might
also be expected to change the thermal expansion. They do, although the difficulty
of making measurements in the vicinity of 1 K means that the range of available
data is limited.

As usual, the most complete data are for vitreous silica. WHITE [4.15] has
measured the coefficient of linear expansion α down to 1.5 K, and his results are
shown in Fig.4.2 and plotted as the corresponding Gruneisen parameter, calculated
using values of heat capacity for the same sample, together with γ_S calculated
from the pressure derivatives of the sound velocities [4.16].

The magnitude of the discrepancy between γ_S and γ_T is best illustrated by the
numerical values at 1.5 K, the lowest temperature for which data are available.
γ_S is -2.2 and γ_T is -17, where γ_T has been calculated from White's data using
(4.7) with $\beta = 3\alpha$. WHITE, in fact, analysed his results differently, dividing both
α and C into the sum of two terms, one proportional to T and the other to T^3. The
two terms give two different Gruneisen constants of about -40 and -7, respectively.
However, this analysis is not completely satisfactory. The T^3 term in the heat
capacity results not only from the Debye contribution calculated from the measured
sound velocities, but also from a term which is probably connected with those excitations responsible for the linear term. If, as an alternative to White's analysis, the total heat capacity is divided into Debye and excess contributions, the

Gruneisen constant appropriate to the additional excitations can be calculated by using the acoustic γ_S for the Debye contribution. The 'excess' γ is then about -50 at 1.5 K.

Data also exist for the organic polymer PMMA [4.17], which shows similar although less marked behaviour. The analysis of the data into Debye and non-Debye contributions is more difficult, partly because γ_S is positive, and partly because the heat capacity is varying much more rapidly with temperature than in SiO_2. However, LYON et al. [4.17] deduced that there is a term in the thermal expansion proportional to temperature, and show that it gives a Gruneisen constant of -7.5. They also argued that a similar value should be found in SiO_2.

The only other detailed measurements below 4 K are by WHITE et al. [4.18], who have measured the thermal expansion and heat capacities between 2 and 30 K of a range of sodium silicate glasses, together with the pressure derivatives of the sound velocities of the same glasses. The results are summarized in Fig.4.3 in the form of γ_T as a function of temperature, and γ_S. Three important points emerge from these results, although all the implications are not yet clear. First, the magnitude of γ decreases as the concentrations of Na_2O is increased up to 20%, with both linear and cubic temperature dependent terms in the expansion coefficient decreasing together. At higher concentrations of Na_2O γ reverses sign and increases in magnitude. Second, the pressure derivatives of the sound velocities show exactly the same trend, changing sign at about 25% Na_2O. Finally, the discrepancy between γ_T and γ_S also changes sign at about the same composition, γ_T remaining roughly twice as large as γ_S, irrespective of sign.

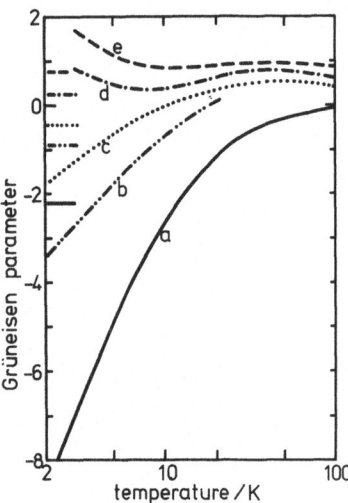

Fig.4.3. The Grüneisen parameter γ_T as a function of temperature for various glasses in the SiO_2/Na_2O system; a) SiO_2, b) 15% Na_2O in SiO_2, c) 20%, d) 30% and e) 40%. The corresponding values of γ_S calculated from the measured sound velocities are also plotted

4.3.2 Discussion

Gruneisen parameters of this magnitude, much larger than typical lattice values of about unity, are not unique to amorphous solids. Measurement of the pressure variation of the infra-red absorption frequency of Li^+ in KBr imply a value for γ of about 50 [4.19], and direct measurements of the thermal expansion of alkali halides containing impurity ions [4.20] lead to Gruneisen constants of +300 for $KCl:Li^+$, +40 for $NaCl:OH^-$ together with a very large and negative γ in $NaCl:CN^-$. The physical origin of the large values of γ is clear in at least some of these examples, and it is useful to examine them before discussing the glasses.

There is convincing experimental evidence that Li^+ in KCl is a substitutional impurity which can occupy one of eight possible sites displaced in the <111> directions [4.21]. By contrast, Li^+ in KBr occupies a single minimum at the centre of the K^+ vacancy. Calculations [4.22] of the local potential seen by Li^+ ion in a K^+ vacancy have shown how this difference can occur. The change in potential energy of Li^+ as it is displaced from the centre of the vacancy in a sensitive balance between repulsive and attractive terms, the latter arising in the main from the polarization induced in the surrounding lattice by Li^+. The repulsive interaction between Li^+ and the neighbouring ions is of course a very rapid function of the separation, while the attractive term varies less rapidly with separation. This difference means that either the attractive or the repulsive terms can dominate at the centre of the site depending upon the size of the unit cell. For a particular host-ion combination the larger the unit cell is, the more likely a non-central minima in the potential, since the attraction between the Li^+ and its neighbours will be more important at the centre of the site. This is illustrated by calculations in $KBr:Li^+$, but, of course, a comparison between Li^+ in KCl and KBr is complicated by the fact that the magnitudes of the repulsive and attractive potentials are different in the two hosts. However, the important point for the purposes of thermal expansion is that the potential at the centre of the site is extremely sensitive to the average interatomic separation and so to pressure; and it is this sensitivity that leads to the large values of Gruneisen parameter in $KBr:Li^+$.

In $KCl:Li^+$ there is a second factor which leads to the even larger values of γ. The ground state of the off-centre ion is split by tunneling between the minima by an amount that depends exponentially both on the height of the potential barrier and the spacial separation between the minima. This exponential dependence compounds the sensitivity of the form of the multiminima potential to pressure, and a rough estimate based on the calculation of QUIGLEY and DAS [4.22] or WILSON et al. [4.23] shows that values of over 100 can be easily obtained for the effective γ of the $KCl:Li^+$ tunnelling state.

SHEARD [4.24] has described how large *negative* values of γ can occur in tunneling systems where the impurity has an intrinsic dipole moment. In this case,

even if the centre of mass of the impurity sits at the centre of the site, there
will be a number of potential minima corresponding to different orientations of the
dipole moment with respect to the surrounding lattice. The barrier to rotation
from one minimum to another will clearly increase when pressure is applied to the
solid, leading to a decrease in the tunnel splitting and hence to a negative γ. When
both rotation and off-centre motion occur, the resulting γ is presumably smaller
in magnitude than in systems where only one is present. NaCl:OH is believed to be
in this mixed category, and does indeed have a γ much smaller than KCl:Li$^+$.

Tunneling states, similar to although not as precisely specified as those in
alkali halides, have been proposed as an explanation for the additional excitation
that appears in the heat capacity [4.25,26]. As described elsewhere in this book,
the presence of such states can explain a wide range of thermal, acoustic and
dielectric data in glasses and, at first sight, appears to explain the large negative values of γ. However, there are important differences between tunneling states
in a crystal and a glass which can be clarified by examining NaCl:CN$^-$. This tunneling system differs from the ones mentioned earlier. The tunnel splitting is
small, and the energy difference between the ground and first excited states is
governed by random strains which perturb the six minima. NaCl:CN$^-$ is an almost
purely rotational tunnelling system, but although it shows a negative γ this cannot be for the reasons outlined above. The tunnel splitting gives a very small contribution to the energy and so the effect of changes in the tunnel splitting will
be small.

Similar arguments apply to tunneling states in a glass. The energy of a tunneling state depends on two parameters, the tunnel splitting and the asymmetry,
both of which will be distributed over a wide range of values. A given energy can,
therefore, arise almost entirely from the tunnel splitting or, alternatively, from
the asymmetry. The effect of this is similar to that seen in NaCl:CN$^-$, where the
large γ associated with the tunnel splitting is reduced. A numerical estimate of
this effect, based on reasonable distributions of tunnel splitting and asymmetry
[4.27] shows that in glasses the effective γ can be reduced by a factor of up to
50.

It is clear that the most important effect comes from the asymmetry. In general terms it is easy to see how this might arise, although detailed application
to a particular case is not easy and has not been attempted. All tunneling states,
both in crystals and glasses, are strongly coupled to the lattice, with coupling
constants of the order of 1 eV. Since typical energies are about 10^{-4} eV, comparable to kT at 1 K, the fractional change in energy per unit strain is enormous.
Direct use of (4.5) can give γs of up to 10^4. This is true for a single tunneling
state, but in practice the effective γ will be much reduced when the appropriate
averages over all tunneling states are taken into account. However, the large effect of strains on the asymmetry of tunneling states can provide values of γ comparable with those found by experiment.

Turning to Fig.4.3 any detailed explanation of the results in SiO_2/Na_2O glasses in terms of the tunneling (or any other) model presents formidable problems. The sketchy heat capacity data [4.28] show that the density of tunneling states remains roughly constant as the concentration of Na_2O is increased. The smaller discrepancy between γ_T and γ_S in SiO_2 Na_2O as compared to SiO_2 must therefore result from a different microscopic structure of the tunneling states in SiO_2 Na_2O, or different distribution functions for the tunnel splittings and asymmetries.

A speculative idea, related to the model put forward by KLEIN et al. [4.29] to explain the fact that the density of excess states is almost the same in all amorphous solids, is that in SiO_2 Na_2O the tunneling states incorporate Na^+ ions and therefore have large electrical dipole moments. Dipole-dipole interaction will therefore play an important part in determining the asymmetry energy, just as magnetic interactions do in spin glasses [4.26]. Increasing the Na_2O concentration increases the number of tunneling states, but decreasing the average distance between dipoles also increases the spread of asymmetry energies and maintains an almost constant density of states. Further, if dipole-dipole interactions are an important influence on the asymmetry energy, it can be shown fairly easily that the tunneling state γ should approach a value of order unity, in agreement with experiment.

In conclusion, it appears that the tunneling model has sufficient flexibility to explain the limited thermal expansion data that are available for glasses below 4 K. Clearly the explanation is very sensitive to the details of the structure, just as in the case of tunneling defects in alkali halide crystals, and requires a more detailed microscopic understanding of these states than do the other low temperature experiments in glasses. However, the experimental results in glasses differ so markedly from those in crystals that further measurements of thermal expansion at low temperatures should lead to a better understanding of the low frequency excitation in amorphous dielectrics.

References

4.1 T.H.K. Barron: In *Lattice Dynamics*, ed. by R.F. Wallis (Pergamon, New York 1965) p.247
4.2 P.H. Gaskell: Trans. Faraday Soc. *62*, 1493 (1966)
4.3 F.W. Sheard: Philos. Mag. *3*, 1381 (1958)
4.4 O.L. Anderson, G.T. Dienes: In *Non-Crystalline Solids*, ed. by V.D. Frechette (Wiley, New York 1960)
4.5 J.R. Ferraro, M.H. Manghnani: J. Phys. Chem. Glasses *13*, 116 (1972)
4.6 K. Chow, W.A. Phillips, A.I. Bienenstock: In *Phase Transitions 1973*, ed. by H.K. Henisch, R. Roy (Pergamon, New York 1973) p.333
4.7 R.J. Bell, N.F. Bird, P. Dean: J. Phys. C *1*, 299 (1968)
4.8 G.K. White, J.A. Birch, M.H. Magnhnani: J. Non Cryst. Solids *23*, 99 (1977)
4.9 J.R. Ferraro, M.H. Manghnani: J. Appl. Phys. *43*, 4595 (1972)
4.10 A. Makishmer, J.D. Mackenzie: J. Non Cryst. Solids *23*, 305 (1976)

4.11 R.D. McCammon, G.K. White: Phys. Rev. Lett. *10*, 234 (1963)
4.12 G.K. White: Proc. R. Soc. London A*286*, 204 (1965)
4.13 H.J. McSkimin: J. Acoust. Soc. Am. *30*, 314 (1958)
4.14 R.C. Zeller, R.O. Pohl: Phys. Rev. B*4*, 2029 (1971)
4.15 G.K. White: Phys. Rev. Lett. *34*, 204 (1975)
4.16 C.J. Kurkjian, J.T. Krause, H.J. McSkimin, P. Andreatch, T.B. Bateman: In *Amorphous Materials*, ed. by R.M. Douglas, B. Ellis (Wiley, New York 1972) p.463
4.17 K.G. Lyon, G.L. Salinger, C.A. Swenson: Phys. Rev. B*19*, 4231 (1979)
4.18 G.K. White, J.A. Birch, M.H. Manghnani: J. Non-Cryst. Solids *23*, 99 (1977)
4.19 I.G. Nolt, A.J. Sieves: Phys. Rev. *174*, 1004 (1968)
4.20 C.R. Case, K.O. McClean, C.A. Swenson, G.K. White: AIP Conf. Proc. *3*, 183 (1972)
4.21 V. Narayanamurti, R.O. Pohl: Rev. Mod. Phys. *42*, 201 (1970)
4.22 R.J. Quigley, T.P. Das: Phys. Rev. *177*, 1340 (1969)
4.23 W.D. Wilson, R.D. Hatcher, G.J. Dienes, R. Smoluchowski: Phys. Rev. *161*, 888 (1967)
4.24 F.W. Sheard: In *Thermal Expansion 1971*, ed. by M.G. Graham, H.E. Hagy, AIP Conf. 1972, p.155
4.25 P.W. Anderson, B.I. Halperin, C.M. Varma: Philos. Mag. *25*, 1 (1972)
4.26 W.A. Phillips: J. Low Temp. Phys. *7*, 351 (1972)
4.27 W.A. Phillips: J. Low Temp. Phys. *11*, 757 (1973)
4.28 R.B. Stephens: Phys. Rev. B*13*, 852 (1976)
4.29 M.W. Klein, B. Fischer, A.C. Anderson, P.J. Anthony: Phys. Rev. B*18*, 5887 (1978)

Additional References

G.J. Morgan, G.K. White, J.G. Collins: "Thermal Expansion of Amorphous As_2S_3 at Low Temperatures", Phil. Mag. (to be published)
T.H.K. Barron, J.G. Collins, G.K. White: "Thermal Expansion of Solids at Low Temperatures", Adv. Phys. *29*, 626 (1980)

5. Thermal Conductivity

A. C. Anderson

With 5 Figures

There exists an excellent fundamental understanding of nonelectronic thermal transport in crystalline dielectrics and metals at low temperatures. Our perception of
thermal transport in amorphous materials is, however, at a much more primitive
level. This chapter reviews the experimental evidence which suggests that phonons
transport heat in glassy materials, then presents the frequency dependence of the
phonon mean free path as derived empirically. The magnitude and frequency dependence of this mean free path appear to be characteristic of noncrystalline materials.
The several phonon-scattering mechanisms proposed to explain the mean free path
are discussed; that which is most attractive at present (for temperatures $\lesssim 10$ K)
is based on the concept of localized excitations which are related, perhaps, to
quantum mechanical tunneling. The chapter then looks at several attempts to study
the localized excitations using thermal conductivity measurements. The reader is
reminded that parallel studies have made use of ultrasonic, specific heat, and
other measurement techniques which are described elsewhere in this book.

5.1 Thermal Transport in Crystalline Materials

We first review those aspects of thermal transport in crystalline materials which
will be of particular use in the ensuing discussion. In a crystal, heat is trans-
ported by phonons which are the elementary excitations of the system. The thermal
conductivity can be expressed as

$$\varkappa(T) = (1/3) \sum_i \int_0^{\omega_D} C_i(\omega)v_i(\omega)\ell_i(\omega)d\omega \tag{5.1}$$

for an isotropic material where the sum is over the three phonon modes, v and ℓ
are the phonon velocity and mean free path, and $C(\omega)$ is the contribution to the
lattice specific heat of those phonons having frequency ω. Integration is from
$\omega = 0$ to the maximum (Debye) frequency ω_D. For a more detailed discussion of this
calculation see ZIMAN [5.1] or KLEMENS [5.2].

   The typical temperature dependence of $\varkappa(T)$ for a crystalline material is shown
by curve I in Fig.5.1. Above ≈ 10 K phonons are scattered by intrinsic processes

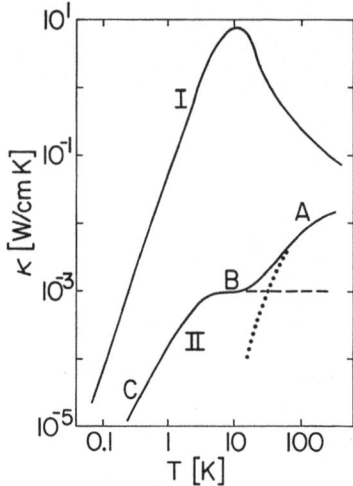

Fig.5.1. Thermal conductivity $\varkappa$ of crystalline quartz (I) and fused quartz (II). The thermal conductivities of all other amorphous materials which have been studied have a temperature dependence and magnitude similar to that of II

such as phonon-phonon and umklapp interactions. Below ≈ 10 K the phonon mean free path is often limited by nonspecular scattering from the surfaces of a sample. Nevertheless, $\varkappa$ is generally defined as the total heat flux divided by the cross-sectional area of the sample so that the "thermal conductivity" becomes size dependent. The portion of the curve below 10 K in Fig.5.1 is depressed with decreasing sample dimensions.

Discussion of phonon thermal transport can be greatly simplified by use of the dominant-phonon approximation in which an average phonon frequency $\nu(= \omega/2\pi)$ is related to temperature T by $\nu \approx 10^{11}T$ [Hz/K]. In this approximation

$$\varkappa = (1/3)C\bar{v}\ell \tag{5.2}$$

or, in the Debye limit of no dispersion,

$$\varkappa = 4.08 \times 10^{10}T^3\ell/\bar{v}^2 \text{ [W/cmK]} . \tag{5.3}$$

Here C is the specific heat per unit volume contributed by phonons and $\bar{v}^2 = (1/3)$ $(v_\ell^{-2} + 2v_t^{-2})^{-1}$, where v_ℓ and v_t are the longitudinal and transverse sound velocities. It must be emphasized that the dominant-phonon approximation and (5.2,3) are useful only if ℓ is weakly frequency dependent. For example, Rayleigh scattering with $\ell \propto \omega^{-4}$ does *not* by itself produce $\varkappa \propto T^{-1}$ as is occasionally stated in the literature.

There is sufficient confidence in our basic understanding of thermal transport in crystalline materials that measurement of the thermal conductivity has become a widely employed spectrometric technique with which to investigate the interaction of phonons with electrons, magnetic excitations, lattice defects, impurities, and phonons. One obtains the interaction strength from ℓ, having deduced ℓ from (5.1-3) using measured values of $\varkappa$ and v. It is of course a broad-band spectrometric technique; details in ℓ or $\kappa(T)$ other than those caused by phase transitions are not resolved if their width ΔT is less than $\approx T$.

5.2 Thermal Transport in Amorphous Materials

The thermal conductivity technique described above was naturally extended to a study of amorphous materials, but the results were, initially, sufficiently opaque that one could not be certain if phonons were responsible for carrying the thermal energy. The typical temperature dependence of $\kappa(T)$ for amorphous materials is shown in Fig.5.1 as curve II. Region A of that curve decreases slowly with decreasing temperature as observed by BERMAN [5.3]. Near ≈ 10 K the thermal conductivity is only weakly temperature dependent, and hence region B is often referred to as a "plateau". Region C was measured by ANDERSON et al. [5.4], who found κ to decrease roughly as T^2. The behaviour of curve II was noted to be typical of glasses, plastics, and even frozen greases; the magnitude and temperature dependence of κ appeared to depend on the amorphous structure of the materials rather than the chemical composition. This fact was utilized to estimate the low-temperature thermal conductivities of many construction materials in the absence of empirical data; yet this remarkable property of glassy materials attracted little attention until 1971 when ZELLER and POHL [5.5] published a paper containing careful measurements of κ in the T^2 region. They suggested that the $\approx T^2$ behavior might be associated with another unusual property of glassy materials, namely a specific heat at low temperatures in excess of that expected for phonons and having a temperature dependence of the form T^m, $m \approx 1$. This paper catalyzed considerable theoretical and experimental work on the low temperature properties of glassy materials.

The first experimental investigations were directed to a determination of the mechanism of heat transfer in glassy materials. Although phonons were known to propagate in glasses from acoustic and light scattering (Brillouin) measurements such as those of FLUBACHER et al. [5.6], it was not obvious whether phonons were responsible for heat transport. Other mechanisms have been suggested; see TANTTILA [5.7] and references cited therein.

It was expected that the role of phonons could be deduced from thermal conductivity measurements in which the dimensions of the sample were reduced sufficiently that scattering of phonons from the surfaces of the sample would dominate all other scattering processes. The experimental results could then be compared with (5.3) in which the acoustic velocities v_ℓ and v_t were known, and for which ℓ was obtained from the dimensions of the sample. ZAITLIN et al. [5.8] found it rather difficult to produce nonspecular boundary scattering in amorphous materials. In crystalline materials nonspecular scattering is caused primarily by defects at the surfaces; these defects do not exist in glasses. Nevertheless, evidence for boundary

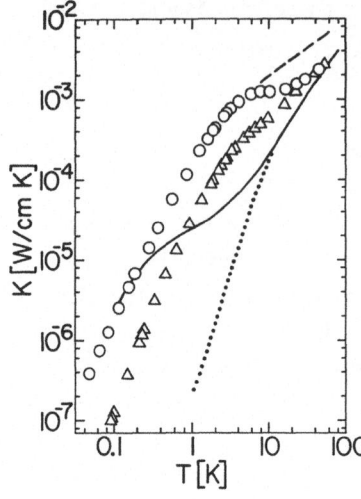

Fig.5.2. Thermal conductivity $\varkappa$ of silicate glasses; data of [5.10]. O-bulk or intrinsic thermal conductivity; Δ-reduced thermal conductivity caused by holes intentionally placed in the sample, the mean free path between holes being 5×10^{-4} cm. The solid line represents the measured thermal conductivity of porous Vycor, from [5.11,12]. The dashed and dotted curves are explained in the text

scattering of phonons was observed by POHL et al. [5.9] in thin glass fibers which had been roughened by a chemical etchant. More definitive results were obtained by ZAITLIN and ANDERSON [5.10] from thin plates of either glass or plastic which contained microscopic holes oriented perpendicular both to the surfaces of the plates and to the direction of heat flow. Scattering was caused by the holes, which could be varied in both size and density to adjust the boundary scattering mean free path. The results for one sample are shown in Fig.5.2. Below 0.2 K the results agree with (5.3) to 10% for a factor of 10 variation in the measured ℓ. This agreement suggests that i) thermal phonons alone transport heat, ii) ℓ is essentially the same for both transverse and longitudinal phonons, and iii) there exist additional states or excitations which contribute the excess T^m term to the specific heat and which are localized, that is they do not contribute to heat transport.

5.2.1 The Phonon Mean Free Path

The data of Fig.5.2 also contain useful information in the region above 0.2 K. The holes introduced into the samples caused a decrease in $\varkappa$ at all temperatures up to ≈ 20 K; notably the plateau was lowered. This behaviour can be explained most readily if the phonon mean free path of a glass is highly frequency dependent as shown in Fig.5.3. For simplicity consider the portion of the curve labelled B to be vertical at frequency ω_0. For $\omega > \omega_0$, ℓ is frequency independent; for $\omega < \omega_0$, $\ell \propto \omega^{-1}$. Phonons of frequency $\omega > \omega_0$ would provide a thermal conductivity shown by region A of Fig.5.1, plus the dotted extension. Phonons of $\omega < \omega_0$ would produce region C where $\varkappa \propto T^2$, plus region B and the dashed extension. The total thermal conductivity is the sum of these two individual curves. The width of the plateau is determined by the magnitude of curve A relative to curve B-C. A wide plateau is produced by a small relative magnitude of A; the plateau could be

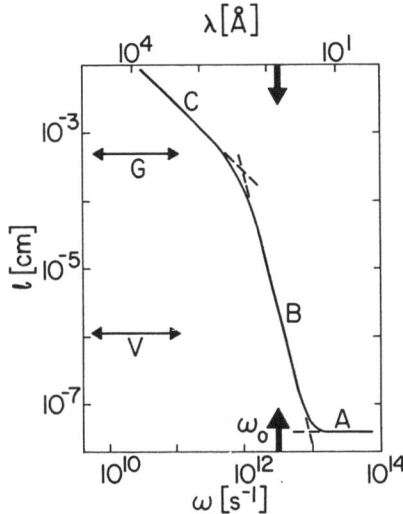

Fig.5.3. Proposed frequency dependence of the intrinsic phonon mean free path ℓ in glassy materials. This curve is quantitatively appropriate for the borosilicate glass of Fig. 5.2. Region C is drawn with a slope of -1. Region B at ω_0 is drawn with a slope of -4, however this slope could be larger in magnitude. G and V indicate the frequency-independent mean free paths introduced by the holes in the borosilicate glass and the porous Vycor, respectively, if diffraction about the holes at small ω is neglected

nearly absent for a large relative magnitude of A. A more detailed discussion of these curves may be found in [5.10].

If the qualitative frequency dependence shown in Fig.5.3 is correct, amorphous materials behave as a low-pass filter. Primarily long-wavelength, low-frequency phonons carry the heat even at temperatures as high as 20 K. These phonons, if in equilibrium, would be characteristic of a temperature near 2 K. The curve also indicates that spectral diffusion is slow—that low-frequency and high-frequency phonons interact only weakly. Thus, in transient behavior, portions of the phonon population may be driven far out of equilibrium from other sectors of the frequency spectrum. Finally, in calculating the thermal conductivity, (5.1) must be used at least for $1 \lesssim T \lesssim 50$ K. The dominant-phonon approximation is not appropriate because of the strong frequency dependence of ℓ.

There are three tests which can help determine if the frequency dependence of Fig.5.3 is realistic. The first is the insertion of holes as already discussed. The holes provide a constant, frequency independent mean free path denoted by line G on Fig.5.3. No phonon can have an ℓ greater than this value, and hence phonons of $\omega \lesssim 5 \times 10^{11}$/s experience a reduced, constant mean free path. As a result, the regions C and B of Fig.5.1 would be lowered, and region A would not be changed. This is as observed in Fig.5.2. Also the region C would attain a temperature dependence of T^3 rather than T^2, also as observed.

If the holes introduced were of sufficiently small diameter d, diffraction of phonons should occur. For a phonon wavelength $\lambda < 5d$ boundary scattering would dominate, but for $\lambda > 5d$ a Rayleigh-like scattering proportional to ω^{-3} would dominate [5.10]. Vycor glass contains a myriad of holes having d ≈ 60 Å. Its thermal conductivity, measured by STEPHENS [5.11] and by TAIT [5.12], is shown in Fig.5.2

as the solid curve. The frequency independent phonon mean free path that would be produced by these holes in the absence of diffraction is indicated by line V on Fig.5.3. The calculated thermal conductivity using line V is shown as the dotted curve on Fig.5.2. Indeed, the measured conductivity agrees with the calculation above ≈ 10 K, but below 10 K the measured conductivity rises above that calculated assuming boundary scattering from the holes since diffraction becomes increasingly important. At 1 K the difference is a factor of 100. Eventually, below 0.2 K, the intrinsic scattering processes of the bulk glass again produce a $\approx T^2$ dependence as though the holes were not present.

Evidence of the commencement of diffraction in Vycor is obvious near ≈ 10 K in Fig.5.2. This corresponds to $\lambda \approx 30$ Å in the dominant-phonon approximation. Yet diffraction is not expected unless $\lambda \gtrsim 300$ Å. In brief, the strong influence of diffraction at temperatures as high as 10 K in Vycor provides a qualitative indication that the phonons involved in heat transport near 10 K are of long wavelength and low frequency. A quantitative calculation [5.10] using the information in Fig.5.3 and allowing for diffraction agrees well with the data.

Additional evidence that Fig.5.3 is qualitatively correct is obtained from measurements of the thermal boundary resistance R_B between dissimilar materials. This thermal resistance arises from the reflection and refraction of thermal phonons at the interface, and is well understood [5.13]. However, measurements made involving the interfaces between epoxy resins and various metals gave values of R_B a factor of 100 larger than calculated as shown in Fig.5.4 [5.14-16]. This behavior apparently arises because the epoxy is amorphous and a low-pass filter; only low-frequency phonons can contribute to thermal transport across the interface. The result of a calculation by MATSUMOTO et al. [5.16] incorporating this fact, shown by the solid curve in Fig.5.4, is in good agreement with the experimental data.

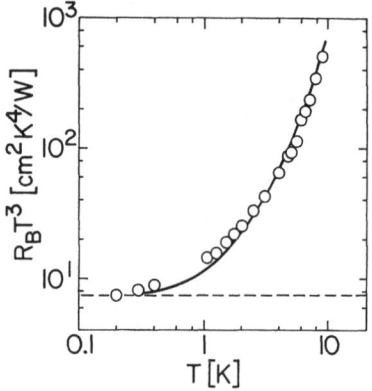

Fig.5.4. The apparent thermal boundary resistance R_B at an interface between copper and epoxy, multiplied by T^3. Data of [5.16]. The dashed line is calculated from standard theory. The solid line uses the same theory but includes the effect that essentially only phonons of frequency $\omega < \omega_0$ (Fig.5.3) participate in heat exchange across the interface. No adjustable parameters were used in the calculation of either curve

There is other, less direct evidence that Fig.5.3 depicts the correct behavior
of ℓ. ROTH and ANDERSON [5.17] have observed that, at low temperatures, the crystallites in partially crystallized materials scatter phonons much as the holes discussed previously. This scattering, plus that intrinsic to glasses as represented
by Fig.5.3, adequately accounts for thermal transport in at least some partially
crystallized and phase separated materials [5.18,19], and perhaps also in compo-
sites consisting of powders in an amorphous matrix [5.14,20,21].

We therefore have considerable evidence that phonons transport heat in glassy
materials, and that their mean free path behaves *qualitatively* as depicted in Fig.
5.3. The question now arises as to the mechanism or mechanisms responsible for
phonon scattering.

5.2.2 Phonon Scattering Mechanisms

A region of nearly constant phonon mean free path, such as A in Fig.5.3, was suggested as early as 1949 by KITTEL [5.22] to explain the behavior of region A in
Fig.5.1. Using measured values of $\varkappa$, C, and $\bar{v}$ in (5.2) required ℓ to be essentially
constant at a value of $D \approx 6$ Å for several glasses. The dimension D was assumed to
be associated with some correlation length. There appears to have been no attempt
to determine if $D \approx 6$ Å is realistic from the viewpoint of phonon propagation and
phonon scattering. An indication of the problems encountered can be seen by making
the naive assumption that D is the average size of crystallites. Then ℓ constant
at a value $D \approx 6$ Å would be reasonable for phonon wavelengths $\lambda \lesssim D/3$ since "grain-
boundary" scattering would occur at the interfaces between crystallites. At frequencies for which $\lambda \gtrsim D/3$ the scattering process should begin a transition to dif-
fractive or Rayleigh scattering as observed by MASON and McSKIMIN [5.23] in acoustic measurements, and ℓ would increase with decreasing frequency. However, the thermal conductivity data cannot be fit if the constant mean free path at A in Fig.5.3
exists only at frequencies above $\omega = 2\pi v/\lambda \approx 6\pi v/D \approx 10^{14}/s$. In brief, the dimension
of $D \approx 6$ Å seemingly required by $\varkappa$ is not consistent with continuum-wave theory.
This is hardly surprising since in this region phonon wavelengths would be of order
an interatomic distance in the amorphous structure. Indeed, $\ell \approx 6$ Å is the shortest
mean free path which is conceivable from an atomistic point of view.

The "cutoff" at ω_0 (part B) in Fig.5.3 is responsible for the occurrence of the
plateau at B in Fig.5.1. A frequency dependence of $\ell \propto \omega^{-s}$, $s \geq 4$, is required to
fit thermal conductivity data on silicate glasses [5.10] and epoxy resins [5.16].
It was initially suggested by KITTEL [5.22] that this strong frequency dependence
was caused by Rayleigh scattering about the cells of diameter D discussed above.
Alternatively, ZELLER and POHL [5.5] proposed that Rayleigh scattering occurred at
essentially every atom, since each atom was displaced from a crystal-lattice site.
However, using present continuum-wave theory and the known properties of glasses,
ZAITLIN and ANDERSON [5.10] and JÄCKLE [5.24] found it impossible to obtain a suf-

ficiently large magnitude of Rayleigh scattering to explain part B of Figs.5.1 or 5.3. The possibility of scattering by impurities as discussed by SALINGER [5.25] can be eliminated since $\varkappa$ versus T is nearly universal for a variety of amorphous materials, and since little sensitivity to impurities has been observed by DAMON [5.26], LASJAUNIAS et al. [5.27], or others. Phonon scattering from the cores of dislocations as suggested by SCOTT and GILES [5.28] and DUBEY [5.29] seems unlikely. First, there is the question of the existence of dislocations in glasses; second, the required density of dislocations is so large that the dislocation model would not be valid. A modified dislocation or line-defect has been suggested by LU and NELKIN [5.30], but no direct evidence for such defects was observed in small-angle neutron scattering measurements by SPOONER and HASTINGS [5.31].

The explanation for the plateau most widely quoted in the literature is not correct. Because of scattering associated with local fluctuations in acoustic velocity, KLEMENS [5.2,32] thought that longitudinal phonons could account for heat transport in regions B and C of Fig.5.1, and that primarily transverse phonons would account for region A. However, the boundary scattering measurements of Fig. 5.2 provide evidence that transverse phonons dominate the heat transport at low temperatures just as in a crystalline dielectric. The plateau has also been ascribed by KLEMENS [5.2] and by DREYFUS et al. [5.33] to dynamic scattering of phonons by localized excitations, much as observed in crystalline materials [5.34]. This appears a possibility and we shall return to it again.

Scattering in region C was first associated with unspecified "internal boundaries" by CHANG and JONES [5.35]. Somewhat related explanations have utilized long-range correlations in the structural disorder [5.2,36,37]. These models have difficulty in explaining how the $\approx T^2$ behavior can persist over such a broad range of temperatures as observed. Scattering by the long-range strain fields of dislocations, as suggested by SCOTT and GILES [5.28], DUBEY [5.29], LU and NELKIN [5.30], and COUCHMAN et al. [5.38], is not a likely explanation for the same reason as described previously under the discussion of the plateau. A dynamic scattering process, a very rapid mechanical relaxation, has been proposed by FULDE and WAGNER [5.39], but has not been related to any physically measurable parameters. Other models designed to explain the excess specific heat in glasses have not been extended to a discussion of the thermal conductivity, but conceivably could be [5.40-43].

A model which assumes the existence of tunneling sites has been developed by PHILLIPS [5.44] and by ANDERSON et al. [5.45] to explain, initially, both the T^2 dependence of the thermal conductivity in region C of Fig.5.1 as well as the T^m dependence of the excess specific heat. Tunneling sites have been studied at low temperatures in crystals [5.34,46]; they originate when atoms or molecules alter position or orientation via quantum-mechanical tunneling. The result is a ground state and one or more excited states of the material. Generally, in crystals the energies $E = \hbar\omega$ of these excitations are narrowly defined since the barriers through

which tunneling occurs are all similar. The relaxation times of the excited states
in different systems can vary by orders of magnitude [5.46,47]; this is partly
related to the size of the barriers. A long relaxation time implies a weak interaction with phonons. If tunneling sites are intrinsic to amorphous materials, a
broad statistical distribution of barriers would be expected. There would thus be
both a broad spectrum of energies $n(E)$ and of relaxation times. The distribution
$n(E)$ must, however, be only weakly energy dependent at small E to fit both the
$\approx T^2$ dependence of the thermal conductivity and the $\approx T^1$ dependence of the excess
specific heat.

The tunneling-sites model has been very successful in predicting and explaining
the results of acoustic measurements on glasses [5.48-51]. These measurements involve phonons in the range $\approx 10^7 - 10^{10}$ Hz, which thus overlap the range of phonons
involved in thermal conductivity measurements. It is then appropriate to note that
the mean free paths deduced from low-intensity acoustic measurements for both transverse and longitudinal modes are the same as those deduced from thermal conductivity [5.50,52]. Apparently the same scattering mechanism is involved throughout
region C of Fig.5.3, and this mechanism is described very well by the tunneling-
sites model.

In the tunneling-sites model the resonant scattering of phonons by sites of
energy $E = \hbar\omega$ produces a mean free path [5.48,49]

$$\ell_C = (A\, k/\hbar\omega)\, \coth(\hbar\omega/2kT) \quad . \tag{5.4}$$

Inserted into (5.1), this mean free path produces a T^2 dependence in agreement
with the observed behavior of region C in Fig.5.1. But a problem occurs at higher
temperatures. Assume, as an example, that Fig.5.3 is correct, that
$\ell^{-1} = \ell_B^{-1} + \ell_C^{-1}$ if $\ell > \ell_A$, otherwise $\ell = \ell_A$. (Here $\ell_B \propto \omega^{-p}$ with $p \gtrsim 4$ to produce
the "cutoff" at ω_0). Inserting ℓ into (5.1) produces the dashed curve in Fig.5.2;
the plateau is absent. The reason is that, in the tunneling theory, ℓ_C of (5.4)
is also temperature dependent; Fig.5.3 in this case should actually be three dimensional with T as the third coordinate. As T increases at fixed ω, ℓ_C becomes
larger. Thus the low-frequency phonons which, as explained previously, transport
the heat throughout the plateau region have increasing mean free paths with increasing T. Equation (5.4) therefore is *not* compatible with the empirical information contained in Fig.5.3.

But there is an additional scattering process associated with the tunneling
sites, a nonresonant relaxation process [5.48,49]which, acting by itself, would
produce a mean free path $\ell_x = BT^{-3}$. Thus

$$\ell_C = \left\{ [(Ak/\hbar\omega)\, \coth(\hbar\omega/2kT)]^{-1} + [B\, T^{-3}]^{-1} \right\}^{-1} \quad . \tag{5.5}$$

The contribution from $B\,T^{-3}$ prevents the unlimited increase in ℓ_c and allows the theory to be fitted to the plateau [5.10].

There is some cause for concern in applying ℓ_x at frequencies of order 10^{11} Hz. The derivation of ℓ_x assumes a homogeneous medium, whereas in reality the distance between scattering centers or tunneling sites lying within a width $\Delta\omega/\omega \approx 0.1$ is of order the phonon wavelength. The derivation also assumes the relaxation of tunneling sites is rapid relative to $\nu^{-1} \approx 10^{11}$ s. Nevertheless, the value required for ℓ_x to fit the thermal conductivity data on fused quartz was found by ZAITLIN and ANDERSSON [5.10] to agree within 20% with that obtained from direct measurements of the (saturated) acoustic attenuation by JÄCKLE et al. [5.49]. This agreement may be fortuitous, but it does suggest that the calculation of the relaxation mechanism may provide a reasonable approximation, at larger ω and T, of the basically three-phonon processes involving tunneling sites.

The possibility of using the existing calculations at larger ω has led to a suggestion of an alternative cause of the plateau, or part B of Fig.5.3. PICHE et al. [5.53], PELOUS and VACHER [5.54], and GOLDING et al. [5.55] have presented evidence that the density of localized states n(E) is not a constant as initially was assumed in the tunneling-sites model, but has a weak dependence on E. ZAITLIN and ANDERSON [5.56] have shown that if n(E), as deduced from acoustic measurements, is combined with (5.5) in a calculation of the thermal conductivity using (5.1), a rough fit to the experimental data can be obtained with $\ell_B^{-1} = 0$. That is, no ad hoc assumption is needed to produce the "cutoff". A more realistic calculation of this three-phonon effect has been suggested by MAYNARD [5.57], but has yet to be carried out.

There is one additional piece of evidence that dynamic scattering processes are involved in region B (as well as C) of Fig.5.3. LEADBETTER et al. [5.58] have observed a well defined minimum near 15 K in the plateau region for amorphous As_2S_3. Such a minimum in $\varkappa$ requires an explicitly temperature dependent phonon mean free path. The tunneling-sites model can provide the appropriate mean free path [5.10, 57].

5.2.3 Summary

Thus far in this review we have arrived at a highly unsatisfactory understanding of phonon thermal transport in noncrystalline materials. In region A of Figs.5.1 or 5.3 it is not obvious if a phonon description of thermal transport is valid, even though it has been used for $\approx$ 30 years. The mechanism of phonon scattering in region B has not been elucidated, although tunneling sites may be involved. In region C the scattering appears to be by tunneling sites, yet these sites have not been identified in any of the many amorphous systems which have been studied. We are still at the level of phenomenology.

5.3 Probing the Localized Excitations

Thermal conductivity measurements have been used in an attempt to gain some insight into the physical nature of the localized excitations. It was initially suggested by PHILLIPS [5.44] that open structure was essential for tunneling, and hence tunneling sites would not be expected to occur, for example, in amorphous Ge or Si. Yet the thermal conductivity of amorphous Ge as measured by NATH and CHOPRA [5.59] and LOHNEYSEN and STEGLICH [5.60] appears to have the same temperature dependence as curve II in Fig.5.1 and even a reduction in x in the T^2 and plateau regions caused by boundary scattering as in Fig.5.2. The thermal conductivities of several glassy metals have also been measured, since they are atomically close packed. (The difference in density between the glassy and crystalline states of a metal is only $\approx 2\%$, cf. $\approx 20\%$ for fused quartz). In each case, the temperature dependence and magnitude of the phonon thermal conductivity is characteristic of a glass, independent of whether the glassy metal is quenched from the melt [5.61,62], electrodeposited at room temperature [5.62], or vapor deposited at low temperatures [5.63].

Specific heat measurements on vapor deposited films led to a suggestion that covalent bonding was essential to the existence of the localized excitations [5.64, 65]. Yet thermal conductivity measurements on amorphous Pb-Cu [5.63] and Ti-Be-Zr alloys [5.62], which should be nearly free of such bonding, produced the typical "glassy" phonon conductivity curve.

In brief, the universal behavior of the thermal conductivity curve II of Fig. 5.1 applies to all materials yet measured which are believed to be amorphous. It would appear that this temperature dependence of x could be incorporated into the definition of an amorphous material.

There have been attempts to manipulate the localized excitations in glassy systems through the application of large electric fields [5.66,67], large magnetic fields [5.66,68], or large strain fields [5.69]. No measurable effects were observed in the thermal conductivity. These negative results serve to place limits on parameters in the tunneling-sites (or other) model, but these limits are compromised by assumptions made concerning the nature of the excitations.

There has been a search for localized excitations in materials which are not completely amorphous. If the search were successful, it would provide some qualitative indication of the range of environments conducive to the existence of localized excitations. In addition, the simplified environment might permit a first-principles calculation of the states. NATHAN et al. [5.70] studied a mixed crystal of KBr-KI to determine if compositional fluctuations would cause glasslike behavior. The negative result further supports the suggestion that an amorphous structure is required for the existence of the localized excitations. Evidence has been found in superconducting, crystalline alloys (Ti-Nb, Zr-Nb) for apparently glassy behavior which might be associated with a metastable ω phase [5.71,72]. However,

there remains the possibility that the observed behavior ($\varkappa \propto T^2$, $C \propto T$) is instead caused by a small number of normal conduction electrons [5.72,73].
Evidence of glassy behavior has been observed in the crystalline superionic conductor β-alumina by ANTHONY and ANDERSON [5.74]. This material consists of plates of alumina bonded together by oxygen ions. Other ions, such as Na, Li, K, and Rb, may be placed between the plates where their mobility is very high. As β-alumina is cooled, the mobile ions freeze into a disordered array which is one atomic layer thick and therefore truly two dimensional. The thermal conductivity of β-alumina exhibits both a $\approx T^2$ dependence at low temperatures and a plateau (also $C \propto T^m$). The magnitude of $\varkappa$ is sensitive to the ion present, which indicates that the excitations are localized to the conducting plane between the alumina plates. This plane therefore appears to be a two-dimensional glass, in which case two- and three-dimensional glasses both exhibit the same behavior. This material may be a suitable candidate for simplified calculations because much is known about the positions of the ions in β-alumina, and because it is only two dimensional.

However, caution must be used when one finds a material exhibiting glasslike behavior. As an example, glassy or vitreous carbons were observed by KATERBERG and ANDERSON [5.75] to have a thermal conductivity like that of curve II in Fig. 5.1. These carbons also have an excess specific heat with a temperature dependence of $\approx T^1$, and mechanical properties and a physical appearance typical of glasses. The use of the adjectives glassy and vitreous therefore appears logical. Yet the thermal conductivity may be dictated entirely by phonon scattering from a large density of small voids. If any localized excitations are present, they have little influence on the low temperature properties.

In attempting to apply the tunneling-sites model quantitatively, problems arise in that a single density of states n(E) may not fit all measurements for a particular material. It would appear that n(E) could be written, in a first approximation, as

$$n(E) = n_{11}(E) + n_{12}(E) + n_{21}(E) + n_{22}(E) \quad . \tag{5.6}$$

Here $n_{11}(E)$ and $n_{12}(E)$ represent localized excitations intrinsic to the amorphous state with $n_{11}(E)$ having a relatively fast relaxation rate (strong phonon interaction) and $n_{12}(E)$ a slow relaxation rate. Both $n_{11}(E)$ and $n_{12}(E)$ would contribute to a specific heat measurement obtained in a classical calorimeter while the relatively small fraction, $n_{11}(E)$ alone, would contribute to ultrasonic and thermal transport properties [5.51,76]. The terms $n_{21}(E)$ and $n_{22}(E)$ would be associated with impurities, again with a small portion $n_{21}(E)$ having a relatively fast relaxation rate [5.77,78]. Each of the four terms could have a different dependence on E. These additional sets of adjustable parameters assure a reasonable fit between theory and experiment, but further complicate any attempt towards a definitive test of the tunneling-sites model or an identification of the tunneling sites themselves.

To gain some insight into the possible composition of n(E), SMITH et al. [5.79] have measured the specific heat, temperature dependence of ultrasonic velocity, and thermal conductivity of fused quartz and of neutron irradiated fused quartz. Irradiation reduces the total number of localized excitations. This is not through a process of recrystallization within regions experiencing "thermal spikes" [5.80] since a mild heat treatment at 900°C returns the thermal conductivity [5.81], mass density, and acoustic properties [5.82] to that measured prior to irradiation. The irradiation appears to decrease n(E) through a modification in the amorphous structure. The measured change in thermal resistivity $w = \varkappa^{-1}$ caused by the removal of phonon scattering sites is shown by the solid curve in Fig.5.5. The simplest interpretation of these data is to assume that irradiation changes only n(E), and that the "cutoff" near ω_0 in Fig.5.3 is associated with n(E) as suggested in Sect.5.2.2. The dotted curve of Fig.5.5 is the result of a calculation based on the tunneling-sites model using parameters which are consistent with the resonant and relaxation scattering deduced from the velocity measurements (in a manner similar to that utilized in [5.53]), and using a decrease in density of localized states equal to that obtained from the specific heat measurements. No free parameters are available in the computation. Three qualitative conclusions are obtained from the agreement between experiment and theory indicated in Fig.5.5. First, the irradiation-produced change in $\Delta w/w$ at the plateau (≈ 10 K), which is twice as large as in the region where $\varkappa \propto T^2$, is readily explained in terms of tunneling sites. This lends additional support to the suggestion in Sect.5.2.2 that the "cutoff" is caused by localized excitations. Second, the energy dependence of n(E) responsible for the excess specific heat is essentially the same as for the small and unique fraction $n_{11}(E)$ of localized excitations active in limiting thermal transport. Third, the energy dependence of n(E) is somewhat stronger than previously suggested, namely $n(E) \approx a\ E^{0.3} + b\ E^{3.5}$ with a $\approx 10^2 b$. It remains to be seen if these features apply as well to other amorphous materials.

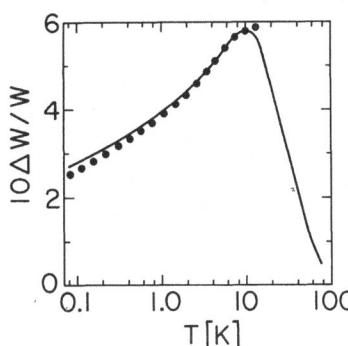

Fig.5.5. Change in the thermal resistivity w of fused quartz caused by neutron irradiation. Data of [5.79]. Solid curve represents the data; dotted curve is calculated using the tunneling-sites model

5.4 Synopsis

The following statements concerning the thermal conductivity of amorphous materials appear to be true in general, but may err in detail. Nonelectronic thermal transport below ≈ 10 K is provided by phonons. The phonon mean free path has a magnitude, frequency, and temperature dependence which is characteristic of the glassy state. Phonon scattering responsible for this mean free path is caused primarily by localized excitations. The most satisfactory description of the excitations presently available, which is also consistent with measurements discussed elsewhere in this book, is a spectrum of sites at which quantum mechanical tunneling occurs.

A number of problems remain to be solved, both theoretically and experimentally. The major problems are the nature and diversity of the localized excitations, and the behavior of the localized excitations at energies above ≈ 1 K ($\approx 10^{11}$ Hz). In addition, a more satisfactory theoretical explanation of thermal conduction in amorphous materials above 10 K is desirable.

Acknowledgments. The author is indebted to P.J. Anthony and J.R. Matey for numerous discussions, and to the National Science Foundation for financial support under contract DMR77-08599.

References

5.1 J.M. Ziman: *Electrons and Phonons* (Oxford, London 1963)
5.2 P.G. Klemens: In *Physics of Non-Crystalline Solids*, ed. by J.A. Prins (North-Holland, Amsterdam 1965) p.162
5.3 R. Berman: Proc. R. Soc. London A*208*, 90 (1951)
5.4 A.C. Anderson, W. Reese, J.C. Wheatley: Rev. Sci. Instrum. *34*, 1386 (1963)
5.5 R.C. Zeller, R.O. Pohl: Phys. Rev. B*4*, 2029 (1971)
5.6 P. Flubacher, A.J. Leadbetter, J.A. Morrison, B.P. Stoicleff: J. Phys. Chem. Solids *12*, 53 (1959)
5.7 W.H. Tanttila: Phys. Rev. Lett. *39*, 554 (1977)
5.8 M.P. Zaitlin, L.M. Scherr, A.C. Anderson: Phys. Rev. B*12*, 4487 (1975)
5.9 R.O. Pohl, W.F. Love, R.B. Stephens: In *Proc. 5th Intern. Conf. on Amorphous and Liquid Semiconductors*, ed. by J. Stuke and W. Brenig (Taylor and Francis, London 1974) p.1121
5.10 M.P. Zaitlin, A.C. Anderson: Phys. Rev. B*12*, 4475 (1975)
5.11 R.B. Stephens: Ph.D. Thesis, Cornell University (1974)
5.12 R.H. Tait: Ph.D. Thesis, Cornell University (1975)
5.13 J.A. Katerberg, C.L. Reynolds, A.C. Anderson: Phys. Rev. B*16*, 673 (1977)
5.14 C. Schmidt: Cyrogenics *15*, 17 (1975); Phys. Rev. B*15*, 4187 (1977)
5.15 F.F.T. de Araujo, H.M. Rosenberg: In *Phonon Scattering in Solids*, ed. by L.J. Challis, V.W. Rampton, A.F.G. Wyatt (Plenum, New York 1976) p.43
5.16 D.S. Matsumoto, C.L. Reynolds, A.C.Anderson: Phys. Rev. B*16*, 3303 (1977); *19*, 4277 (1979)
5.17 E.P. Roth, A.C. Anderson: J. Appl. Phys. *47*, 3644 (1976)
5.18 J.H.A. Laudy: In *Physics of Non-Crystalline Solids*, ed. by J.A. Prins (Wiley, New York 1965) p.189
5.19 F. Canal, M.C. Schmidt, J.P. Redoules, P. Carrara: J. Non-Cryst. Solids *21*, 73 (1976)

5.20 A.C. Anderson, R.B. Rauch: J. Appl. Phys. *41*, 3648 (1970)
5.21 K.W. Garrett, H.M. Rosenberg: J. Phys. D*7*, 1247 (1974)
5.22 C. Kittel: Phys. Rev. *75*, 972 (1949)
5.23 W.P. Mason, H.J. McSkimin: J. Appl. Phys. *19*, 940 (1948)
5.24 J. Jäckle: In *The Physics of Non-Crystalline Solids*, ed. by G.H. Frischat
      (Transtech, Aedermannsdorf 1977) p.568
5.25 G.L. Salinger: In *Amorphous Materials*, ed. by R.W. Douglas, B. Ellis
      (Wiley-Interscience, New York 1970) p.475
5.26 D.H. Damon: Phys. Rev. B*8*, 5860 (1973)
5.27 J.C. Lasjaunias, A. Ravex, M. Vandorpe: Solid State Commun. *17*, 1045 (1975)
5.28 T. Scott, M. Giles: Phys. Rev. Lett. *29*, 642 (1972)
5.29 K.S. Dubey: Solid State Commun. *15*, 875 (1974)
5.30 M.-S. Lu, M. Nelkin: Report 2415, Materials Science Center, Cornell University
      (1975); M.-S. Lu, Ph.D. Thesis, Cornell University (1975)
5.31 S. Spooner, J.B. Hastings: J. Non-Cryst. Solids *22*, 443 (1976)
5.32 P.G. Klemens: Proc. R. Soc. London A*208*, 108 (1951)
5.33 B. Dreyfus, N.C. Fernandez, R. Maynard: Phys. Lett. *26*A, 647 (1968)
5.34 V. Narayanamurti, R.O. Pohl: Rev. Mod. Phys. *42*, 201 (1970)
5.35 G.K. Chang, R.E. Jones: Phys. Rev. *126*, 2055 (1962)
5.36 G.J. Morgan, D. Smith: J. Phys. C*7*, 649 (1974)
5.37 D. Walton: Solid State Commun. *14*, 335 (1974)
5.38 P.R. Couchman, C.L. Reynolds, R.M.J. Cotterill: Nature *264*, 534 (1976)
5.39 P. Fulda, H. Wagner: Phys. Rev. Lett. *27*, 1280 (1971)
5.40 H.B. Rosenstock: J. Non-Cryst. Solids *7*, 123 (1972)
5.41 S. Takeno, M. Goda: Prog. Theor. Phys. *48*, 1468 (1972)
5.42 H.P. Baltes: Solid State Commun. *13*, 225 (1973)
5.43 L.S. Kothari, Usha: J. Non-Cryst. Solids *15*, 347 (1974)
      Usha, L.S. Kothari: Solid State Commun. *15*, 579 (1974)
5.44 W.A. Phillips: J. Low Temp. Phys. *7*, 351 (1972)
5.45 P.W. Anderson, B.J. Halperin, C.M. Varma: Philos. Mag. *25*, 1 (1972)
5.46 G.J. Sellers, A.C. Anderson, H.K. Birnbaum: Phys. Rev. B*10*, 2771 (1974)
      S.G. O'Hara, G.J. Sellers, A.C. Anderson: Phys. Rev. B*10*, 2777 (1974)
5.47 R.O. Rollefson: Phys. Rev. B*5*, 3235 (1972)
5.48 J. Jäckle: Z. Phys. *257*, 212 (1972)
5.49 J. Jäckle, L. Piche, W. Arnold, S. Hunklinger: J. Non-Cryst. Solids *20*,
      365 (1976)
5.50 S. Hunklinger, W. Arnold: In *Physical Acoustics*, Vol.12, ed. by W.P. Mason,
      R.N. Thurston (Academic, New York 1976) p.155, and papers cited therein
5.51 B. Golding, J.E. Graebner: Phys. Rev. Lett. *37*, 852 (1976)
5.52 B. Golding, J.E. Graebner, R.J. Schutz: Phys. Rev. B*14*, 1660 (1976)
5.53 L. Piche, R. Maynard, S. Hunklinger, J. Jäckle: Phys. Rev. Lett. *32*, 1426
      (1974)
5.54 J. Pelous, R. Vacher: Sol. State Commun. *19*, 627 (1976)
5.55 B. Golding, J.E. Graebner, A.B. Kane: Phys. Rev. Lett. *37*, 1248 (1976)
5.56 M.P. Zaitlin, A.C. Anderson: Phys. Status Solidi (b) *71*, 323 (1975)
5.57 R. Maynard: In *Phonon Scattering in Solids*, ed. by L.J. Challis, V.M. Rampton,
      A.F.G. Wyatt (Plenum, New York 1976) p.115
5.58 A.J. Leadbetter, A.P. Jeapes, C.G. Waterfield, R. Maynard: J. Phys. (Paris)
      38, 95 (1977)
5.59 P. Nath, K.L. Chopra: Phys. Rev. B*10*, 3412 (1974)
5.60 H. v. Lohneysen, F. Steglich: Phys. Rev. Lett. *39*, 1420 (1977)
5.61 J.R. Matey, A.C. Anderson: J. Non-Cryst. Solids *23*, 129 (1977)
5.62 J.R. Matey, A.C. Anderson: Phys. Rev. B*16*, 3406 (1977); *17*, 5029 (1978)
5.63 H. v. Lohneysen, F. Steglich: Z. Phys. B*29*, 89 (1977)
5.64 G. Krauss, W. Buckel: Z. Phys. B*20*, 147 (1975)
5.65 A. Comberg, S. Ewert: Z. Phys. B*25*, 173 (1976)
5.66 C.N. Hooker, L.J. Challis: In *Phonon Scattering in Solids* (La Documentation
      Francaise, Paris 1972) p.364
5.67 R.B. Stephens: Phys. Rev. B*14*, 754 (1976)
5.68 R.A. Fisher, G.E. Brodale, E.W. Hornung, W.F. Giauque: Rev. Sci. Instrum. *39*,
      108 (1968)

5.69 T.L. Smith, J.R. Matey, A.C. Anderson: Phys. Chem. Glasses *17*, 214 (1976)
5.70 B.D. Nathan, L.F. Lou, R.H. Tait: Solid State Commun. *19*, 615 (1976)
5.71 L.F. Lou: Solid State Comm. *19*, 335 (1976)
5.72 M. Ikebe, S. Nakagawa, K. Hiraga, Y. Muto: Solid State Commun. *23*, 189 (1977)
5.73 L.F. Lou: Phys. Rev. B*14*, 3914 (1976)
5.74 P.J. Anthony, A.C. Anderson: Phys. Rev. B*14*, 5198 (1976); *16*, 3827 (1977)
5.75 J.A. Katerberg, A.C. Anderson: J. Low Temp. Phys. *30*, 739 (1978)
     J.W. Gardner, A.C. Anderson: J. Appl. Phys. *50*, 3012 (1979)
5.76 J.L. Black, B.I. Halperin: Phys. Rev. B*16*, 2879 (1977)
5.77 S. Hunklinger, L. Piche, J.C. Lasjaunias, K. Dransfeld: J. Phys. C*8*, L423 (1975)
5.78 T.L. Smith, A.C. Anderson: Phys. Rev. B*19*, 4315 (1979)
5.79 T.L. Smith, P.J. Anthony, A.C. Anderson: Phys. Rev. B*17*, 4997 (1978)
5.80 W. Primak: *The Compacted States of Vitreous Silica* (Gordon and Breach, New York 1975)
5.81 A.F. Cohen: J. Appl. Phys. *29*, 591 (1958)
     J.H. Crawford, A.F. Cohen: Bull. Int. Inst. Refrig. *1*, 165 (1958) (These specific heat data appear to be in error by a factor of ≈4)
5.82 R.E. Strakna: Phys. Rev. *123*, 2020 (1961)

Additional References

D.A. Ackerman, D. Moy, R.C. Potter, A.C. Anderson, W.N. Lawless: "The Glassy Behavior of Crystalline Solids at Low Temperatures", Phys. Rev. B (to be published)
W. Dietsche, H. Kinder: "Spectroscopy of Phonon Scattering in Glass", Phys. Rev. Lett. *43*, 1413 (1979)

6. Acoustic and Dielectric Properties of Glasses at Low Temperatures

S. Hunklinger and M. v. Schickfus

With 17 Figures

In the past several years increasing attention has been paid to the properties of amorphous solids at low temperatures. This chapter deals with the elastic and dielectric behavior of these substances.

6.1 General Comments

The elastic and dielectric behavior of amorphous solids at low temperatures differs completely from that of crystalline solids. For example, the acoustic and dielectric absorption is strongly enhanced compared with crystals and a large absorption peak is found around liquid nitrogen temperature in many glasses. Below helium temperature a number of anomalous effects are observed, which can be attributed to low-energy excitations closely related to the amorphous state. These excitations are not only responsible for the anomalously high specific heat but also determine the elastic and dielectric properties. Detailed acoustic and electric measurements have been carried out in order to gain more information on the microscopic nature of these low-energy states.

For simplicity we restrict our discussion mainly to simple inorganic glasses and omit the description of the motion of ions in multicomponent glasses and of structural units like side chains in organic polymers. Furthermore, we consider only frequencies up to 40 GHz and temperatures well below room temperature. We exclude phenomena due to coherent effects at ultralow temperatures which are discussed in detail in Chap.7.

6.2 Acoustic and Dielectric Properties Above 10 K

6.2.1 Absorption

The most prominent and general feature of the ultrasonic behaviour of glasses is an absorption that is strongly enhanced compared to crystals [6.1]. In addition, in many glasses a broad absorption maximum occurs at temperatures around liquid nitrogen (Fig.6.1) [6.2,3]. With increasing frequency the peak moves slowly to higher

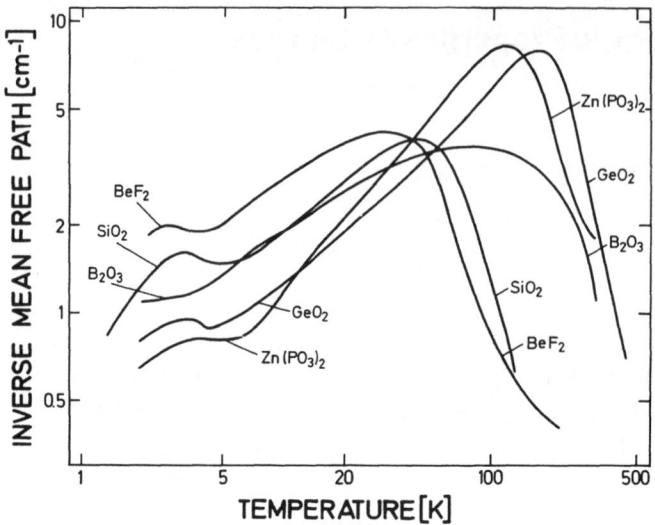

Fig.6.1. Acoustic absorption (or inverse mean free path) of several glasses below room temperature. Data have been taken for longitudinal waves at 20 MHz (after [6.12])

temperatures [6.4-9]. The magnitude of the absorption per wavelength is found to be the same for both longitudinal and transverse polarization [6.7]. While the height of this maximum increases linearly with frequency [6.3-9], a quadratic frequency dependence of the absorption is found on its high temperature side [6.10, 11]. This peak has been found in many glasses, e.g., in glassy SiO_2, GeO_2, BeF_2, As_2O_3, $Zn(PO_3)_2$ [6.7,12], in multicomponent glasses (see, for example, [6.13-16], and in the metallic glasses PdSi [6.17] and Nb_3Ge [6.18]. In some glasses with a low glass transition temperature like As_2S_3, the peak appears only as a shoulder on an absorption rapidly growing with temperature caused by the softening of the structure near the glass transition [6.19,20]. The thermal history of the sample may influence the height, but not the position of the peak [6.21]. On the low-temperature side a second peak or shoulder appears at temperatures below 20 K [6.5,12] which will be discussed in more detail in Sect.6.3.1.

There is a close relationship between the acoustic and the dielectric properties of glasses in this temperature range (Fig.6.2) as pointed out by ANDERSON and BÖMMEL [6.4]. Again, a broad maximum with temperature is the prominent feature of the absorption behavior. Apart from SiO_2 and silica-based multicomponent glasses it has been observed in GeO_2, BeF_2 and As_2S_3 [6.22-29].

In Fig.6.3, the inverse of the temperature at which this maximum occurs in SiO_2 is plotted against the logarithm of the measuring frequency for both dielectric and acoustic measurements. The fact that acoustic and dielectric data fit onto the same plot shows clearly that the same mechanism is responsible for

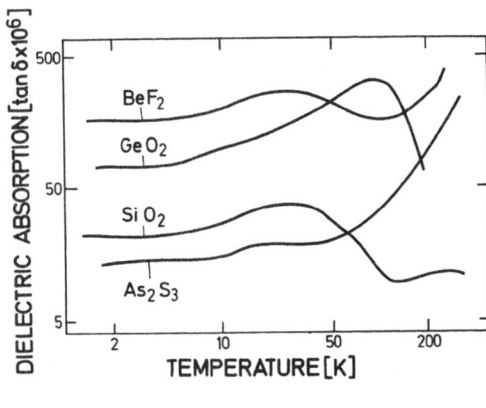

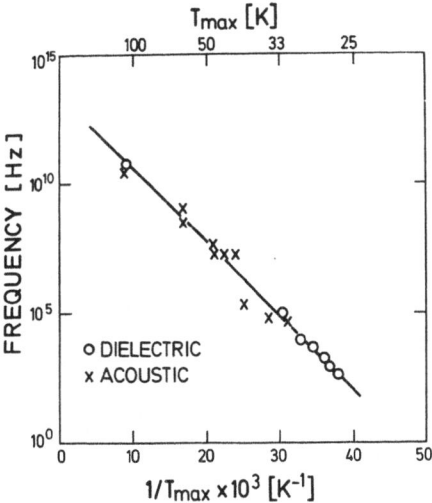

Fig.6.2. Dielectric loss of several glasses below room temperature at 1 kHz (after [6.25,28])

Fig.6.3. Arrhenius plot (frequency versus inverse peak temperature) of the absorption peak observed in vitreous silica in ultrasonic and dielectric measurements

acoustic and dielectric losses. Furthermore, this Arrhenius plot indicates that the origin of the absorption lies in a thermally activated relaxation process [6.4].

The acoustic or electrical absorption caused by such a relaxation is described by the well-known Debye equation

$$\alpha = NA \frac{\omega^2 \tau}{1 + \omega^2 \tau^2} \tag{6.1}$$

where A is the relaxation strength, as discussed below. The relaxation time τ and A give information about the processes taking place. For a thermally activated process the Arrhenius relationship

$$\tau = \tau_0 \exp(V/kT) \tag{6.2}$$

holds, where τ_0 is a constant and V the activation energy[1]. For fused silica, for example, we deduce V/k = 570 K and $\tau_0 \simeq 10^{-13}$ s from Fig.6.3. It turns out, however, that the experimental width of the absorption peak (Fig.6.1) is larger than expected from (6.1) with a single value of V. For a better description a distribution P(V) of the activation energy and consequently of the relaxation time τ has to be assumed. For this purpose we replace the number of intersating centers N by the distribution P(V) and integrate over all energies, i.e., $N \rightarrow \int P(V)dV$. A Gaussian

---

[1] In a more rigorous description the free energy should be used. Omission of the entropy contricution causes a negligible error in our case.

distribution with a width of roughly 400 K has turned out to give a fairly good approximation of the acoustic results [6.30].

The relaxation strength is given by

$$A^a = D^2/4\rho v^3 kT$$

and

$$A^e = \pi\tilde{p}^2/c_0\sqrt{\varepsilon}'kT$$

(6.3)

for the acoustic[2] and electric case, respectively. The deformation potential D describes the energy shift $\delta E = D \cdot e$ of the relaxing states in the strain field e. Here we have assumed that D does not depend on the activation energy, otherwise an appropriate average has to be taken. Similarly, $\delta E = \underline{p} \cdot \underline{F}$ in the electric case, where $\underline{p}$ is the electrical dipole moment of an individual relaxing particle and $\underline{F}$ the electric field strength. $\tilde{p}$ represents an average over all orientations and a possible distribution of $\underline{p}$. From Fig.6.1 we estimate $ND^2 \simeq 10^{-4}$ for SiO_2. If D is of the order of 1 eV (see Sect.6.5.4) we obtain $N \simeq 5 \times 10^{19}$ cm^{-3}, indicating that only a small fraction of atoms contribute to the attenuation process [6.4].

Of course, such a formal analysis is not restricted to fused quartz and can be applied to other glasses as long as the observed maximum is well defined. In many glasses, however, the peak is very broad or degenerated into a shoulder. In this case an analysis in terms of the Debye equation becomes doubtful because unrealistic distributions of the activation energy would have to be assumed. Obviously, other processes of essentially unknown nature enhance the acoustic and dielectric absorption, especially at high temperatures and at very high frequencies [6.16,29].

There is, however, a distinct difference between the acoustic and dielectric behavior (see Fig.6.1,2): Whereas the acoustic attenuation decreases rapidly with decreasing temperatures for T < 40 K, a more or less temperature independent background is observed in the dielectric measurements.

A number of microscopic models have been proposed to explain this relaxation process in pure, simple glasses. The common assumption is the existence of localized structural defects which can exist in two different configurations. For vitreous silica it was proposed by ANDERSON and BÖMMEL [6.4] that a fraction of oxygen atoms are located in a double-well potential and can move from one well to the other by a transverse motion (see defect A in Fig.6.4). Similarly two potential minima, but in the bond direction (defect B), have been proposed by STRAKNA 6.31 . Another possibility is the rotation of SiO_4 tetrahedrons (defect C) by a small angle in a double-well potential [6.32]. These models are equally plausible for other tetrahedrically bonded materials and similar ideas can also be developed for glasses of different structure or even for polymers. Very recently new concepts

---

[2]For simplicity we do not write explicitly the indices defining longitudinal and transverse polarizations of acoustic waves.

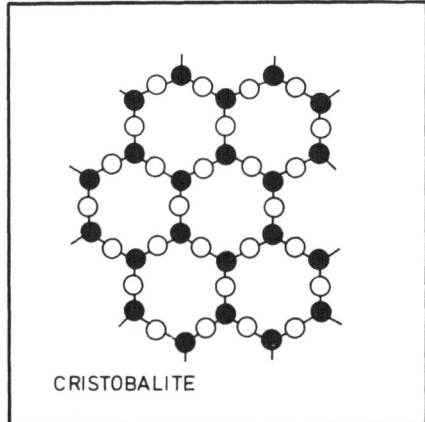 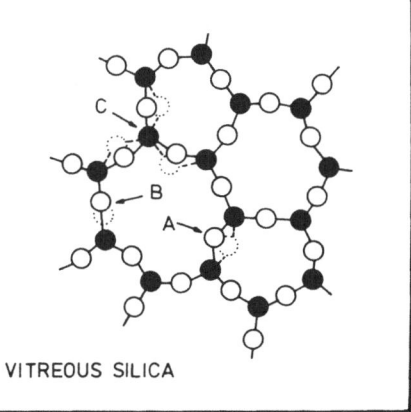

CRISTOBALITE   VITREOUS SILICA

Fig.6.4. Schematic two-dimensional representation of the structure of cristobalite, a crystalline modification of SiO_2, and of vitreous silica (from [6.30]). Full circles represent silicon atoms and open circles oxygen atoms. Three possible types of defects are indicated by arrows

on the nature of defects in amorphous materials have been developed, mainly in order to explain the optical properties of chalcogenides and SiO_2-based glasses (see, for example, [6.33-36]. It is, however, not yet clear whether this type of defect could also play a role in the explanation of the phenomena we are discussing here.

6.2.2 Sound Velocity and Dielectric Constant

In contrast to the weak negative temperature coefficient of the sound velocity in pure dielectric crystals [6.2], much stronger variations, both positive and negative, are found in glasses. At the lowest temperatures the velocity of sound increases first with temperature in all glasses following a very simple logarithmic law. This effect will be discussed in more detail in Sect.6.3.2. After reaching a maximum at a few Kelvin, the velocity decreases again in all glasses. At still higher temperatures, however (typically above 50 K), the chemical composition and structure influence the sound velocity: At room temperature tetrahedrically bonded SiO_2, GeO_2, BeF_2 and $Zn(PO)_3$ have a negative temperature coefficient of the sound velocity, whereas in other glasses like B_2O_3 and As_2S_3 it remains positive (Fig. 6.5) [6.2-4,6,9,12-14,16,37,38].

A more uniform behavior is found for the dielectric constant, although the number of materials investigated is much smaller and no measurements in the MHz range exist, where most of the ultrasonic data were taken. For SiO_2, V_2O_5, GeO_2, BeF_2 and As_2S_3 a steady decrease of the dielectric constant towards lower temperatures has been found from room temperature to at least 10 K (Fig.6.6) [6.22,24, 26-28,39,40]. Again a minimum of the dielectric constant (corresponding to a maxi-

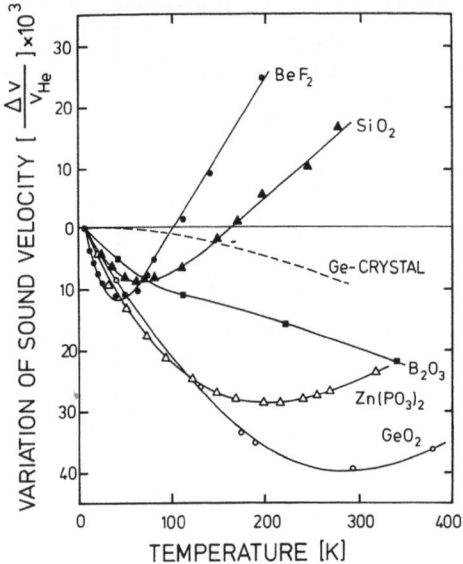

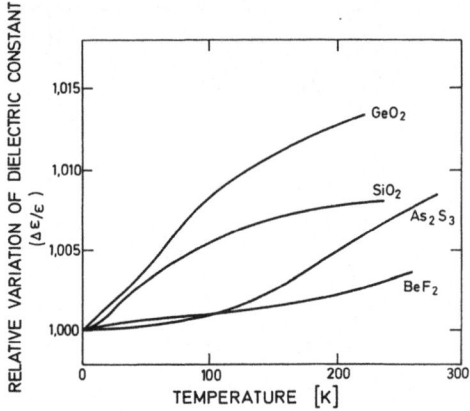

Fig.6.6. Relative variation of the dielectric constant of several glasses below room temperature (after [6.28])

Fig.6.5. Temperature dependence of the velocity of 20 MHz, longitudinal sound waves for several glasses. The velocity of crystalline germanium at 25 MHz is included for comparison. v_{He} is the sound velocity at liquid Helium temperature (from [6.63])

mum of the velocity of light) is found below this temperature. The logarithmic increase following at still lower temperatures will be discussed in Sect.6.3.2.

The common behavior, namely the decrease of the sound velocity v with temperature at temperatures between a few Kelvin and 50 K and the increase of the dielectric constant ε' above 10 K can easily be understood. If we apply the Kramers-Kronig relation to (6.1) we obtain for the relative change of the sound velocity $\Delta v/v$ and of the dielectric constant $\Delta\varepsilon'/\varepsilon'$

$$\frac{\Delta v}{v} = \frac{-A^a v}{2} \int_{V=0}^{V=\infty} P(V) \frac{dV}{1 + \omega^2 \tau(V)^2}$$

$$\frac{\Delta\varepsilon'}{\varepsilon'} = \frac{A^e c_0}{\sqrt{\varepsilon'}} \int_{V=0}^{V=\infty} P(V) \frac{dV}{1 + \omega^2 \tau(V)^2} \quad .$$

(6.4)

Good agreement with the measured values can be achieved [6.4] with the same parameters used to fit the absorption data. In a more sophisticated description, however, not only the contribution of the large absorption peak, but also the low temperature behavior has to be taken into account, but no quantitative analysis covering the whole temperature range has been performed to our knowledge.

The origin of the linear increase of the sound velocity at higher temperatures in vitreous silica or other simple glasses is much less clear. Since a similar effect is not observed in the dielectric constant its origin has to be purely elastic. Several attempts to explain these phenomena are reported in literature, two of which we want to mention here: KULBITSKAYA et al. [6.37,41] attributed the positive temperature coefficient to frozen-in fluctuations of the density and the elastic moduli. According to THOMAS [6.42] it is a peculiarity of tetrahedrially coordinated glasses. He expected an anomalous temperature dependence of the sound velocity for all SiO_2-like glasses with a bond angle at the oxygen atom (or other bridging atoms) which is considerably larger than $90^\circ C$.

6.3 Acoustic and Dielectric Properties Below 10 K

Below 10 K all amorphous substances (probably with the exception of Si and Ge [6.42a,b]) investigated so far show similar anomalous thermal and acoustical pro- perties. These anomalies are attributed to low-energy configurational defects which are thought to be a consequence of the amorphous state.

6.3.1 Acoustic and Dielectric Absorption

a) *Relaxation Effects*

As can be seen from Fig.6.1, additional features appear on the low-temperature side of the relaxation peak discussed in Sect.6.2.1: There is a shoulder or a small additional peak around 5 K followed by a steep decrease of the absorption at still lower temperatures [6.5,8,12]. The position of this low-temperature absorption peak depends more strongly on frequency than that of the high-temperature peak; in SiO_2, only a weak shoulder remains at 1 GHz in contrast to a clearly distin- guishable peak at 20 MHz (Fig.6.7).

Below the maximum the acoustic attenuation drops rapidly as T^3 and becomes in- dependent of frequency [6.15,43]. The peak has been observed in all glasses in- vestigated with the exception of B_2O_3 [6.12] and of glassy metals [6.44,45]. In the case of B_2O_3 it is not yet clear whether the peak would appear at frequencies or temperatures lower than those applied in previous experiments. In amorphous metals the peak or a shoulder has only been observed near T_c in the superconducting alloy $Pd_{30}Zr_{70}$ [6.45a]. In all other cases the absorption below 10 K decreases more weakly than in dielectric glasses and remains frequency dependent.

A behavior corresponding to that of the ultrasonic absorption can be found in the dielectric absorption [6.23,25,46]. Figure 6.8 shows a strong increase of the absorption below 10 K where obviously the absorption maximum was not reached at the lowest temperature of 1.3 K at a frequency of 1 kHz. An important difference that should be pointed out is that the magnitude of the ultrasonic absorption and of the dielectric high-temperature peak is only weakly influenced by the impurity

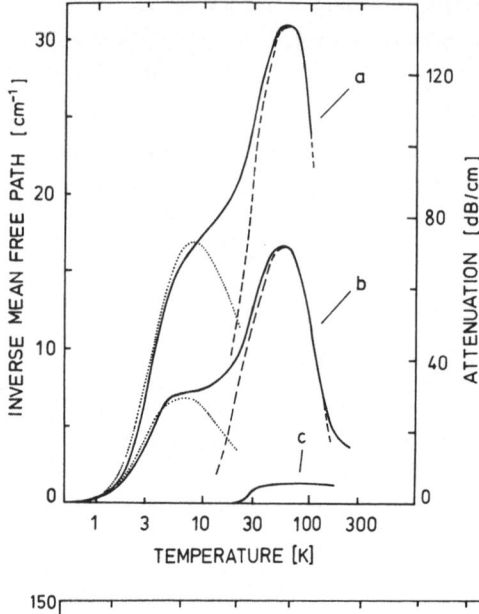

Fig.6.7. Temperature dependence of the longitudinal ultrasonic absorption in vitreous silica at (a) 930 MHz and (b) 507 MHz. Curve (c) represents the absorption in crystalline quartz at 1000 MHz for comparison. Dashed and dotted lines are fits to the relaxation processes discussed in Sects. 6.2.1,6.4.2 (from [6.63])

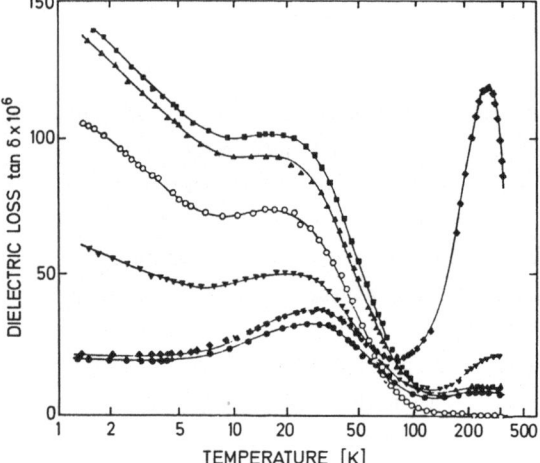

Fig.6.8. Dielectric loss at 1 kHz in vitreous silica of varying OH-content. Note that the height of the peak occurring around 30 K does not depend on the OH-content in contrast to the absorption at lower temperatures (■ Spectrosil B, 1190 ppm OH; ▲ vitreous silica, 1160 ppm OH; o Corning 7940, 940 ppm OH; ▼ Vitreosil OG, 430 ppm OH; ◆ Infrasil, 30 ppm OH; ● Spectrosil WF, 20 ppm OH) (after [6.25]). The maximum occurring above room temperature is probably due to additonal impurities

content of the material. The magnitude of the dielectric low-temperature peak, however, depends strongly on the presence of OH-groups or, more generally, on polar impurities.

b) *Resonant Interaction*

The first comparison of the phonon lifetimes in vitreous silica calculated from thermal conductivity with those deduced from light scattering [6.47] revealed a large discrepancy. The intense phonon pulses generated in the light scattering experiment were much less attenuated than the thermal phonons of the same frequency.

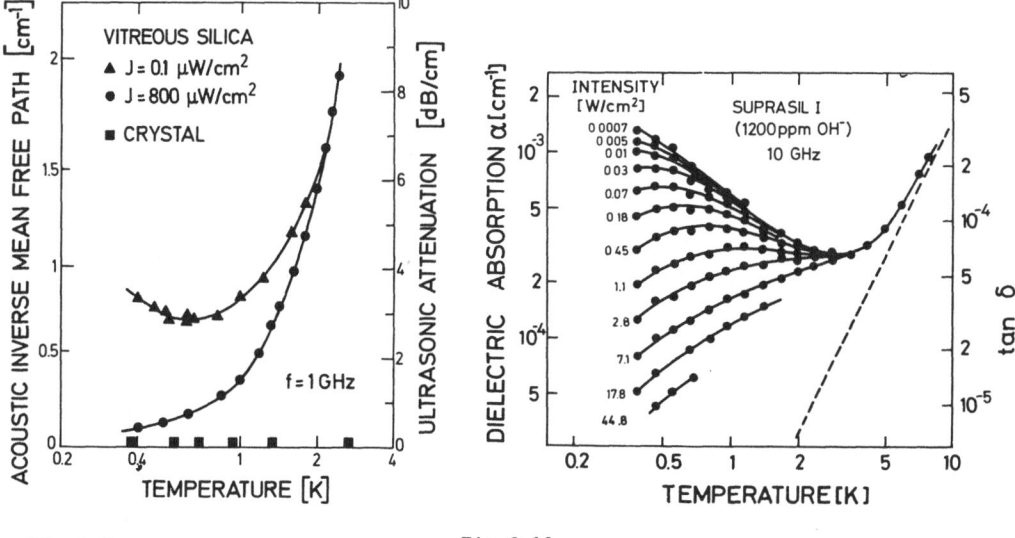

Fig.6.9   Fig.6.10

Fig.6.9. Temperature dependence of the ultrasonic absorption in vitreous silica for longitudinal waves of 1 GHz. At higher intensities a continuous decrease with temperature is observed, whereas at low intensities the absorption rises again below 0.7 K. The absorption of a quartz crystal is included for comparison (from [6.53])

Fig.6.10. Temperature dependence of the dielectric absorption in vitreous silica Suprasil I (1200 ppm OH) at 10 GHz for different microwave intensities. The dashed line indicates the contribution of the relaxation process (from [6.54])

Subsequent ultrasonic experiments between 0.5 and 10 GHz and at varying intensi-ties have shown that a new absorption mechanism dominates at very low temperatures [6.43,48-52]. Figure 6.9 shows the two important consequences of this process: on cooling, the absorption decreases continuously at *high acoustic intensity* whereas at *low intensity* the absorption increases again below an absorption minimum around 0.6 K at 1 GHz. Obviously, this absorption process can be "saturated". If the un-saturable contribution to the absorption is subtracted from the total absorption, it turns out that at low intensity the absorption is proportional to $\omega \tanh(\hbar\omega/2kT)$ [6.50].

Recently the corresponding effects in the dielectric absorption of glasses have been observed in vitreous silica and selenium [6.54,55]. Figure 6.10 shows the di-electric absorption coefficient of vitreous silica Suprasil I at 10 GHz and at different intensities. Below 1.5 K and at an intensity below 10^{-3} W/cm^2, the $\tanh(\hbar\omega/2kT)$ dependence of the absorption is again recognizable. Above this tempera-ture the tail of the low-temperature peak (which occurs at a higher temperature at this frequency) dominates the absorption behavior.

The intensity dependence of the absorption in the dielectric case is shown in more detail in Fig.6.11. Above a "critical" intensity, the absorption coefficient

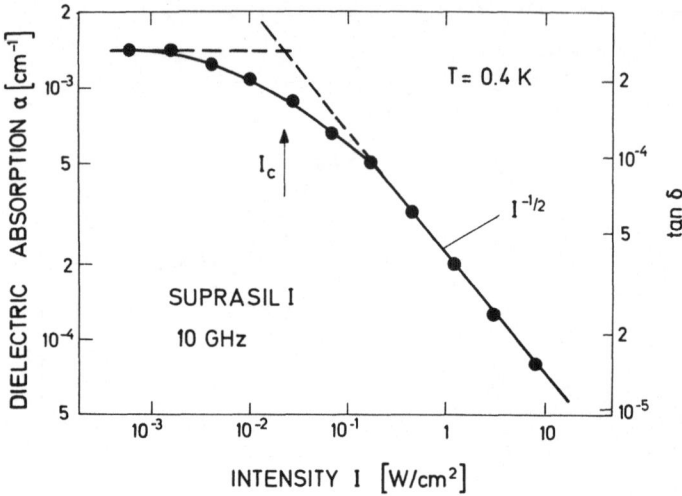

Fig.6.11. Intensity dependence of the dielectric absorption in vitreous silica Suprasil I at 10 GHz and 0.4 K. The critical intensity I_c (Sect.6.4.2) is indicated by an arrow. At higher intensities the absorption becomes proportional to $1/\sqrt{I}$ (from [6.54])

decreases with the square root of the electric field intensity. The first studies of this intensity dependence of the absorption, however, were performed by acoustic measurements in the borosilicate glass BK7 and in pure vitreous silica [6.43,48-52]. In amorphous metals like PdSi a saturation effect has been observed quite recently at rather high intensities [6.56,57]. At very low temperatures (typically below 50 mK), the intensity dependence of the absorption in amorphous dielectrics is even stronger. In Fig.6.12 this effect is shown for longitudinal and transverse sound waves in vitreous silica at 23 mK [6.50].

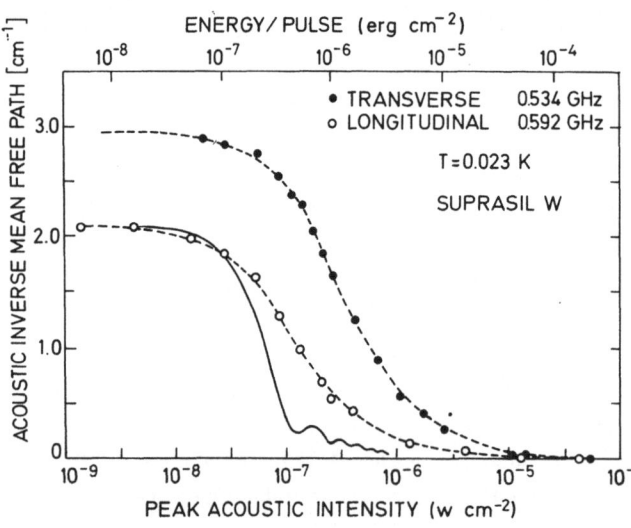

Fig.6.12. Intensity dependence of the ultrasonic absorption in vitreous silica Suprasil W at 23 mK and frequencies around 550 MHz. At this temperature the absorption decreases faster with increasing intensity than $1/\sqrt{I}$ (after [6.50]). The solid line represents a theoretical fit (see Sect. 6.4.2)

6.3.2 Sound Velocity and Dielectric Constant

Along with the absorption, the sound velocity and dielectric constant show variations with temperature which are large compared with those of pure crystals. Below 50 K the increase of velocity with decreasing temperature (Fig.6.5) continues until a frequency dependent maximum is reached (Fig.6.13). At 30 MHz this maximum occurs at 1.5 K in vitreous silica [6.58,59] and is shifted to 6 K at 33 GHz [6.60, 61]. At still lower temperatures a relative variation of the sound velocity $\Delta v/v$ with the logarithm of temperature is observed, until at $T \simeq \hbar\omega/2k$ a shallow minimum is reached [6.62,63] (Fig.6.14). The logarithmic temperature dependence has been observed in oxide glasses, chalcogenide glasses [6.64], amorphous elements like Se [6.65], and metallic glasses [6.57,66,67].[3]

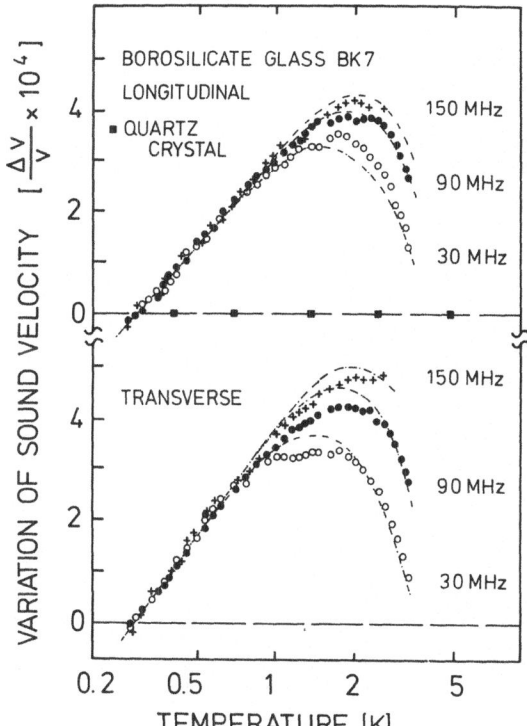

Fig.6.13. Relative variaton of sound velocity with temperature in the borosilicate glass BK7. Below 1 K the temperature dependence becomes logarithmic (from [6.63])

---

[3]It should be pointed out, however, that in $Ge_{10}Si_{12}As_{30}Te_{48}$ the sound velocity seems to be temperature independent below 5 K [6.68].

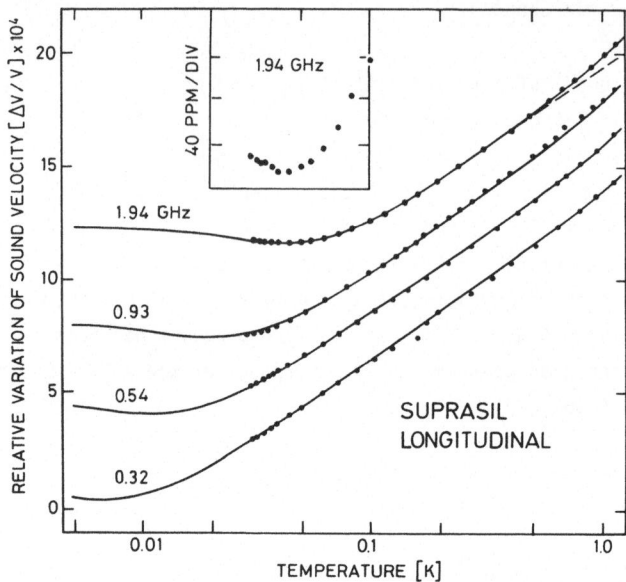

Fig.6.14. Relative variation of sound velocity with temperature in vitreous silica Suprasil at very low temperatures. At $\hbar\omega \simeq 2kT$ a minimum is observed which is discussed in Sect.6.4.3 (after [6.62])

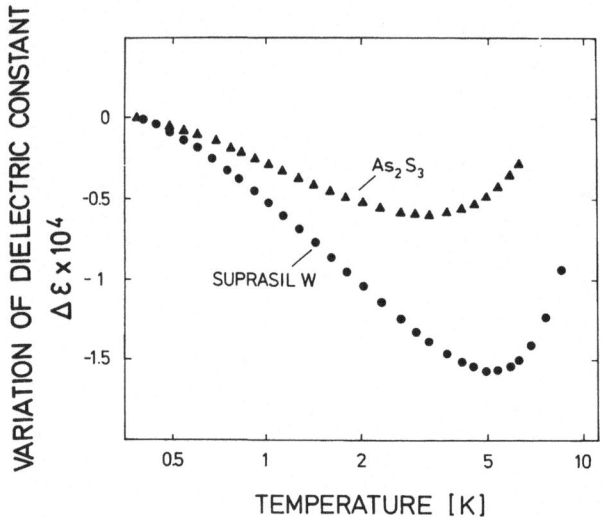

Fig.6.15. Variation of dielectric constant with temperature in vitreous silica Suprasil W (< 5 ppm OH) and vitreous As_2S_3 at 10 GHz (after [6.55])

A completely analogous behavior is found for the relative variation of the dielectric constant of glasses. After a rather steep decrease on cooling, a logarithmic increase is found below a temperature of a few Kelvin at GHz frequencies (Fig.6.15). Such a variation of the dielectric constant has again been observed for a variety of amorphous insulators including polymers like PMMA (Polymethylmetacrylate) and PET (Polyethyleneterephthalate) [6.69,70]. Like the dielectric absorption, the magnitude of this temperature variation of the dielectric constant depends on the concentration of polar impurities [6.71]. Recently the range of these measurements has been expanded to temperatures as low as 2.4 mK in the kHz frequency range [6.46,72]. Although these results qualitatively agree with those reported above, differences occur in the absolute values.

6.3.3 Acoustic Dielectric "Cross"-Experiments

In order to find out whether the excitations responsible for the dielectric and the acoustic low-temperature anomalies are identical, so-called cross experiments have been carried out. In these experiments either the acoustic absorption is measured in the presence of a microwave field of nearly the same frequency, or vice versa. In both cases the results indicate that the same excitations are responsible for both kinds of absorption. A microwave field strong enough to saturate the dielectric absorption (see Sect.6.3.1b) will also lower the ultrasonic absorption [6.73], and strong pulses of ultrasound lower the absorption of weak microwave fields when the ultrasonic wave travels through the portion of the sample exposed to the microwave field [6.74]. This effect can even be used as a sensitive detector for coherent and noncoherent ultrasound [6.75].

6.4 Theoretical Description of the Acoustic and Dielectric Properties by Two-Level Systems

From specific heat measurements [6.76-78] it is known that, in addition to Debye phonons, low-energy excitations are present in amorphous materials (see Chap.3). They can be described in terms of two-level systems with an approximately constant density of states. Their interaction with phonons and microwave photons is the subject of this section.

The dynamics of the two-level systems are characterized by two relaxation times T_1 and T_2. T_1 is the radiative lifetime and is given by [6.15,79]

$$T_1^{-1} = \left(\frac{M_\ell^2}{v_\ell^5} + \frac{2M_t^2}{v_t^5} \right) \frac{E^3}{2\pi\rho\hbar^4} \coth \frac{E}{2kT} \tag{6.5}$$

if only the interaction with the phonon bath is taken into account[4]. Here E is the energy splitting of the two-level system and M_ℓ and M_t represent the deformation potential for the longitudinal and transverse phonon, respectively. T_2 is the

---

[4] In metallic glasses the two-level systems interact not only with phonons but also with electrons. This leads to a considerable enhancement of the relaxation rate T_1^{-1} [6.79].

homogeneous lifetime and is in most cases determined by the interaction between individual two-level systems. This interaction is of elastic nature [6.80,81] and gives rise to spectral diffusion as well [6.82,83]. As a result of the spectral diffusion the number of two-level systems interacting with a perturbing wave depends on time. The only consequence of this phenomenon important here is its influence on the magnitude of the absorption at higher intensities (see Sect.6.4.1). A more detailed concept will be given in the following chapter.

6.4.1 Dynamical Properties of Two-Level Systems

The Hamiltonian of a two-level system in an elastic strain field[5] e is given by (see, for example, [6.15])

$$H = H_0 + H^a = \frac{1}{2}\begin{pmatrix} E & 0 \\ 0 & -E \end{pmatrix} - \frac{1}{2}\begin{pmatrix} D & 2M \\ 2M & -D \end{pmatrix} e \quad . \tag{6.6}$$

Here H_0 represents the two-level system without perturbation and H^a describes the interaction with the sound wave. The deformation potentials D and M have already been defined in (6.3,5).

For a discussion of the dielectric properties H^a has to be replaced by H^e [6.48].

$$H^e = -\frac{1}{2}\begin{pmatrix} \mu & 2\mu' \\ 2\mu' & -\mu \end{pmatrix} E \tag{6.7}$$

Here μ and μ' are the permanent and induced dipole moment of the two-level system, respectively.

There is a formal equivalence of the Hamiltonian of (6.6,7) and that of a spin $\frac{1}{2}$-system in a magnetic field (see, for example, [6.84]). Therefore, the dynamics of the two-level systems are described by the Bloch equations, which were first derived for the description of nuclear magnetic resonance. These equations read as follows [6.84]

$$\dot{X} = -X/T_2 - (E + De)Y/\hbar$$

$$\dot{Y} = (E + De)X/\hbar - Y/T_2 - \frac{2Me}{\hbar} Z \tag{6.8}$$

$$\dot{Z} = \frac{2Me}{\hbar} Y - (Z - Z_e)/T_1 \quad .$$

In magnetic resonance experiments the observed magnetization is proportional to X, Y and Z which represent the expectation values of the spin components. In an ensemble

---

[5]For simplicity we neglect the tensorial character of the strain field and of the deformation potentials.

of spin $\frac{1}{2}$-systems they can vary between 0 and $\pm\frac{1}{2}$. In our case, X and Y reflect the induced elastic or electric polarization. The Z-component, however, can only formally be interpreted as a polarization component (see, for example, [6.85]), since the level splitting is not due to an externally applied static field. The difference in the occupation number between the lower and upper state is $-2Z$; Z_e is the value of Z in its "instantaneous equilibrium", i.e.,

$$Z_e = -\frac{1}{2} \tanh[(E + De)/2kT] \quad . \tag{6.9}$$

In first-order approximation we obtain

$$Z_e = Z_0 + De \frac{dZ_0}{dE} \tag{6.10}$$

where $Z_0 = -\frac{1}{2} \tanh(E/2kT)$ is proportional to the population difference in thermal equilibrium.

In the presence of an oscillatory field $e = e_0 \cos(\omega t)$ the solutions of (6.8) show oscillatory behavior. For pulsed fields of duration τ_p only approximate solutions can be found. Here we consider two cases, namely $\tau_p \gg T_1, T_2$ and $\tau_p \ll T_1, T_2$.

a) Under the condition $\tau_p \gg T_1, T_2$, the "steady state" solution, all transient exponentials have decayed at the end of the applied pulse. With $E = \hbar\omega_0$, we can write for the amplitude $\hat{X}$ of X to a good approximation [6.63,84]

$$\hat{X} = \frac{Me}{\hbar} (Z_0 + \bar{Z}) \left[\frac{1}{(\omega_0 - \omega) - iT_2^{-1}} + \frac{1}{(\omega_0 + \omega) + iT_2^{-1}} \right] \tag{6.11}$$

where $\bar{Z}$ is time independent and is given by

$$\bar{Z} = \frac{-M^2 e_0^2 T_1 T_2 Z_0/\hbar^2}{1 + (\omega_0 - \omega)^2 T_2^2 + M^2 e_0^2 T_1 T_2/\hbar^2} \quad . \tag{6.12}$$

For the amplitude of Z we find the following approximate result

$$\hat{Z} = - De \frac{dZ_0}{dE} \frac{1 + i\omega T_1}{1 + \omega^2 T_1^2} \quad . \tag{6.13}$$

b) In the case where the coupling between the two-level systems and the amorphous network can be neglected (i.e., $\tau_p \ll T_1, T_2$), we have a situation which is equivalent to the case of the free precession in spin experiments [6.84]. We deduce

$$\hat{X} = \frac{Me}{\hbar} Z_0 \left(\left\{ \frac{(\omega_0 - \omega)}{a^2} [1 - \cos(at)] + \frac{i}{a} \sin(at) \right\} \right.$$

$$\left. + \left\{ \frac{(\omega_0 + \omega)}{a^2 b} [1 - \cos(at)] - \frac{i}{ab} \sin(at) \right\} \right) \tag{6.14}$$

where $a = \sqrt{(\omega_0 - \omega)^2 + M^2 e_0^2/\hbar^2}$ and $b = 4\hbar^2 M^{-2} e_0^{-2} \omega\omega_0 - 1$.

It is worth mentioning that the value of X and hence the absorption as well as the dispersion should now be time dependent. The experimentally observed mean values should therefore depend not only on the strength of the perturbing field but also on τ_p, the pulse duration. The appropriate parameter will be the "pulse area" $\theta = \tau_p \cdot Me_0/\hbar$.

In order to determine the influence of the two-level systems on the acoustic properties we calculate their contribution δU to the internal energy U of the amorphous material [6.15]

$$\delta U = -N(2Me,0,De) \cdot (X,Y,Z) \tag{6.15}$$

where N is the number of two-level systems per unit volume. Their contribution to the dynamical elastic modulus is then given by

$$\delta C = \frac{1}{e} \frac{\partial(\delta U)}{\partial e} = -\frac{N}{e}(2MX + DZ) \quad . \tag{6.16}$$

Sound dispersion and absorption are finally obtained from the real and imaginary parts of the elastic modulus, respectively,

$$\delta v(\omega) = (1/2\rho v) \, Re\{\delta C(\omega)\}$$

$$\alpha(\omega) = -(\omega/\rho v^3) Im\{\delta C(\omega)\} \tag{6.17}$$

where v is the velocity of sound without interaction with the two-level sytems.

Because of the similarity between H^a and H^e we can use the acoustic formulae for the description of electrical experiments if we apply the following substitution: $D \rightarrow \mu$; $M \rightarrow \mu'$; $\rho v^3 \rightarrow c\sqrt{\epsilon'}/4\pi$ (in going from acoustic to electrical absorption) and $\rho v^2 \rightarrow -8\pi$ (in going from sound velocity to the real part of the dielectric constant).

6.4.2 Absorption Due to a Distribution of Two-Level Systems

Because of the energy distribution of the two-level systems we have to integrate over all level splittings. Even if a constant density of states n_a and n_e for the acoustically and electrically active systems is assumed, the complexity of the equation requires numerical integration. Analytic expressions can only be found if approximations are introduced.

According to (6.16) we can distinguish between two different contributions to the absorption. Firstly, the X-component of the polarization (6.11) leads to an absorption of resonant character. Secondly, the occupation of the two states is modulated (6.13) and results in an attenuation which has the typical form of a relaxation process. In order to evaluate the resonant part we insert (6.11) into (6.16) and integrate over all energy splittings. The dominant contribution to the integral is due to two-level systems with an energy splitting $E \simeq \hbar\omega$. We obtain

$$\alpha_{res}^a = \frac{\pi n_a M^2 \omega}{\rho v^3} \frac{\tanh(\hbar\omega/2kT)}{\sqrt{1 + I/I_c}} = \frac{\alpha_{cw}^a}{\sqrt{1 + I/I_c}} \tag{6.18}$$

where α_{cw}^a is the absorption in the limit of very small acoustic intensities I. The absorption becomes intensity dependent ("saturated") as soon as $I \gtrsim I_c$, where I_c is a critical intensity defined as

$$I_c = \hbar^2 \rho v^3 / 2M^2 T_1 T_2 \quad . \tag{6.19}$$

The second term of (6.16) describes a relaxation process [6.86] similar to the thermally activated relaxation process discussed in Sect.6.2.1. After perturbation the two-level systems relax into the new thermal equilibrium via the one-phonon process described by (6.5). All thermally populated two-level systems take part in this process, leading to an absorption which cannot be saturated. Although the integration over all energies cannot be carried out analytically in general, we can find an expression in the limit of very low temperatures, where $\omega T_1 \gg 1$. Then the relaxation absorption is given by

$$\alpha_{rel} = \frac{k^3 T^3 \zeta(4, 0.5)}{32\pi\rho^2 \hbar^4 v^3} n_a D^2 \left(\frac{M_\ell^2}{v_\ell^5} + \frac{2M_t^2}{v_t^5} \right) \tag{6.20}$$

where $\zeta(4, 0.5) = 97.38$ is the Riemann zeta function. For the electrical absorption we have to take into account the fact that the relaxation of the two-level systems back to thermal equilibrium after perturbation by an electromagnetic wave is probably purely elastic [6.71]. Therefore, we only have to replace $n_a D^2 / \rho v^3$ by the appropriate electrical quantity $4\pi n_e \mu^2 / c\sqrt{\varepsilon}$ whereas the relaxation times remain unchanged.

At extremely low temperatures (6.14) holds for the resonant contribution, leading to a time dependent absorption. Integration over all level splittings results in

$$\alpha_{res}^a(t) = \alpha_{cw}^a \, J_0 \left(\frac{Me}{\hbar} \cdot t \right) \tag{6.21}$$

where J_0 is the zero order Bessel function. The mean value of the absorption averaged over the pulse duration is

$$\alpha_{res}^a = \alpha_{cw}^a \frac{1}{\theta} \int_{x=0}^{x=\theta} J_0(x) dx \tag{6.22}$$

where θ is the pulse area defined previously. Here we want to point out that this result is only valid for square pulses and that pulse reshaping is not taken into account [6.87]. The observed absorption decreases with increasing intensity. Considering the experimental curves (see Fig.6.12), this effect is superficially very similar to the saturation effect described by (6.18).

6.4.3 Variation of Sound Velocity and Dielectric Constant

The interaction with the two-level systems not only causes an attenuation but also leads to a variation of the velocities of sound and of light. This is described by the real parts of (6.16). The saturation of the absorption has only a negligible influence on this effect since a broad spectral range of two-level systems contributes. Therefore, we set $e = 0$ and we find for the steady state limit [6.63]

$$\delta v = \frac{n_a M^2}{\rho v} \left[Re \left\{ \Psi \left(\frac{1}{2} + \frac{\hbar T_2^{-1}}{2\pi kT} + \frac{\hbar\omega}{2\pi ikT} \right) \right\} - \ln \frac{E_{max}}{2\pi kT} \right] \tag{6.23}$$

where Ψ represents the digamma function (see, for example, [6.88]) and E_{max} is the cutoff of the density of states at high energies. Since $\hbar/T_2 \ll kT$ we can neglect the second term in the argument of the Ψ-function.

In Fig.6.16, the variation of the velocity of sound is plotted as a function of temperature for different frequencies. For $\hbar\omega \ll kT$ the Ψ-function becomes constant and we obtain the following simple expression if we consider the relative variation only [6.58]

$$v(T) - v(T_0) = \frac{n_a M^2}{\rho v} \ln(T/T_0) \tag{6.24}$$

where T_0 is an arbitrary reference temperature. Similarly, a logarithmic frequency dependence is expected for $\hbar\omega \gg kT$. At ultralow temperatures, where T_1, $T_2 \gg \tau_p$ the expression for the sound velocity is more complicated (6.14), since, in addition, the pulse area has to be considered. But because of the broad distribution of the level splittings, the cosine term in (6.14) does not contribute to the integral. Therefore, the variation of the sound velocity as well as of the dielectric constant is not influenced by the occurrence of coherence.

The relaxation process leads to a variation of the sound velocity as well [6.58]. The factor dZ_0/dE (6.13) causes a dominant contribution from systems with an energy near $3 kT$. In contrast to the resonant interaction, the relaxation process leads to a decrease of the velocity with increasing temperatures. At very low temperatures its contribution is hidden by the resonant process but it becomes dominant above a few degrees. Unfortunately, in this temperature range the condition $\omega T_1 \gg 1$ does not hold and, therefore, the contribution of the relaxation process can only be calculated numerically.

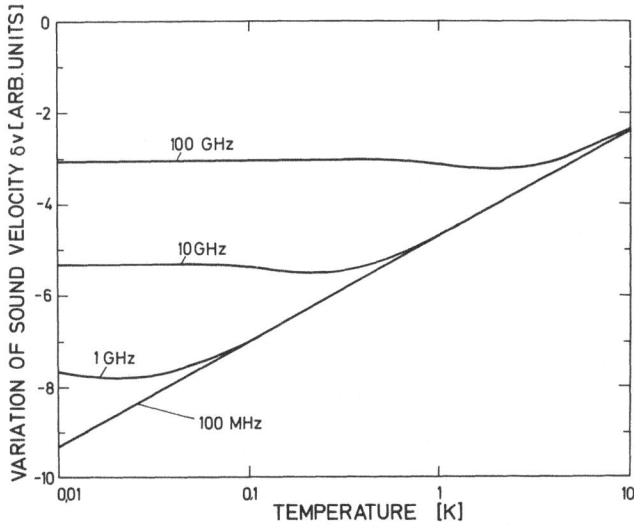

Fig.6.16. Variation of sound velocity with temperature for different frequencies calculated with (6.23). A minimum occurs at $\hbar\omega \simeq 2kT$. At higher temperatures the variation becomes frequency independent and proportional to log T. Below the minimum it is temperature independent, but varies with log ω

6.5 Comparison Between Theory and Experiment

The overall agreement between experiment and theory is surprisingly good for amorphous dielectrics. At higher temperatures (T > 0.1 K) the frequency, temperature and intensity dependences predicted by (6.18,22,24) have been confirmed experimentally. Nevertheless, the question arises whether these steady state solutions are adequate. Measurements on the dielectric behavior have been carried out by quasi-continuous methods [6.55] so that steady state conditions were always fulfilled. In acoustic experiments, however, the pulse duration is typically of the order of 1 μs. Based on saturation recovery experiments it has been argued [6.63] that for measurements around 1 GHz the steady state limit would be adequate down to temperatures of 0.5 K. Very recent experimental and theoretical work [6.89] has resulted in relatively large values for the relaxation time T_1 and consequently this limit has not been reached in most of the experiments. Nevertheless, qualitatively the phenomena remain unchanged. A new analysis has to be carried out, however, for the frequency and temperature dependence of the critical intensity, where the agreement between experiment and theory is rather poor so far [6.63]. In glassy metals the description by (6.20) fails completely even at temperatures as low as 30 mK. We shall discuss this exception in Sect.6.6.

No direct comparison between experiment and theory has been performed for the data obtained at ultralow temperatures where T_1, $T_2 \gg \tau_p$. In Fig.6.12 we have included the theoretical curve deduced from (6.22). Although there is qualitative agreement, a clear deviation from the predicted intensity dependence of the absorption is observed. Probably it is due to the fact that pulse reshaping [6.87] and the distribution of the coupling parameters are not taken into account.

From the magnitude of the measured quantities the mean values of the macroscopic coupling parameters, i.e., the density of states multiplied by the square of the deformation potentials or of the electrical dipole moments, can be deduced. From acoustic measurements $n_a M_\alpha^2$ and $n_a D_\alpha^2$ (α denotes the polarization of the sound wave) can be determined. Although $n_a M_\alpha^2$ is known for a variety of substances, no clear relationship between its magnitude and the chemical composition of the amorphous substance can be recognized. For dielectric glasses $n_a M_\alpha^2$ is of the order of 10^8 erg/cm^2 [6.63], but is considerably smaller in glassy metals [6.57,65,67]. Except for CoP [6.66] the coupling is stronger to longitudinal than to transverse waves. The value of the diagonal coupling parameter $n_a D_\alpha^2$ is known only for a few substances. It is found to be larger than $n_a M_\alpha^2$ by a factor of roughly two or three. A lower limit of the magnitude of M_α and D_α can be estimated by using the density of states as derived from specific heat measurements. Large values, namely deformation potentials of the order of 1 eV, are found. Direct information on M_α is obtained by echo experiments as described in Chapt.7.

Similarly values of $n_e \mu^2$ and $n_e \mu'^2$ can be deduced from dielectric measurements. The strongest coupling is observed for the borosilicate glass BK7, where $6 \cdot 10^{-3}$ and $3 \cdot 10^{-4}$ have been measured, respectively [6.55]. Dipole moments of the order of a few Debyes can be estimated if, again, the density of states from thermal measurements is taken for n_e. There is, however, an important difference between the acoustic and dielectric properties of amorphous materials at low temperatures: In the elastic case the coupling parameter is only modestly influenced by the content of impurities [6.90]. In addition, it is even not very sensitive to the chemical composition of the substance, at least in insulating glasses. In contrast to this the macroscopic dielectric coupling parameters $n_e \mu^2$ and $n_e \mu'^2$ are mainly determined by the impurity content [6.55,71]. This result can be explained if we assume that the "intrinsic" dipole moment of a two-level system can be enhanced by carrying either additional electrical dipoles or charges due to impurities. On the other hand, the assumption is also possible that additional two-level systems with similar elastic properties, but with a much higher dipole moment, are created by the impurities. Both interpretations are supported by the cross-experiment (see Sect.6.3.3). It shows that even in the very "impure" borosilicate glass BK7 not all acoustically excited two-level systems couple strongly to electric fields.

6.6 Microscopic Description: Tunneling Model

Phenomenologically the acoustic and dielectric measurements can be explained by
the existence of two-level systems with a nearly constant density of states. The
tunneling model [6.91,92] offers a microscopic picture for them. As already discussed in previous parts of this book, it is likely that in an amorphous network
certain atoms or groups of atoms can occupy two different configurational states
(see Fig.6.17), i.e., particles of still unspecified nature are able to move in a
double-well potential. At low temperatures these particles cannot jump over the
barrier separating the two wells, but tunneling through the barrier is still possible. Such a tunneling motion has been investigated theoretically [6.79] and experimentally (see, for example, [6.93]) in crystals.

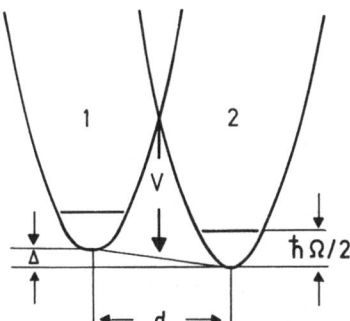

Fig.6.17. Double-well potential with barrier
height V, asymmetry Δ, and distance d between the
two minima. $\hbar\Omega/2$ is the ground state energy

The tunneling motion leads to an energy splitting of the ground state
$\Delta_0 = \hbar\Omega e^{-\lambda}$, where Ω is the vibrational frequency of the particle in an isolated
well. The tunneling parameter λ is given by $\lambda = d\sqrt{2mV}/\hbar^2$, where d is the separation
between the two wells, m is the mass of the tunneling particle, and V the barrier
height. In a random network the environment of the particle can be different at the
two sites, leading to an energy difference Δ between the depths of the wells (see
Fig.6.17). The energy splitting E of the two eigenstates which represent the two-
level system is then given by $E = \sqrt{\Delta^2 + \Delta_0^2}$. The elastic strain field e of long wavelength phonons or the alternating electric field F modulates Δ and Δ_0. The strain
field causes changes in the environment of the tunneling particle, resulting in a
modulation $\delta\Delta$ of Δ. In the electrical case such a modulation is brought about by
the electric field either by the displacement of a charge or by the reorientation
of a permanent dipole with respect to the field direction. Therefore, a relation
exists between the deformation potentials D and M and the dipole moments μ and μ'
or (6.6,17).

$$D = 2\gamma\Delta/E \qquad M = \gamma\Delta_0/E$$
$$\mu = 2p\Delta/E \qquad \mu'= p\Delta_0/E \tag{6.25}$$

where $2\gamma = (d\Delta/de)$ describes the variation of the asymmetry in the depth of the wells with applied strain. Similarly, $2p = (d\Delta/dF)$ represents the effective dipole moment of the tunneling system in the field direction. The parameters Δ and Δ_0 are assumed to be independent variables. In order to compare theory and experiment their distribution must be known. So far none of these functions has been determined experimentally. The generally accepted assumption is that $P(\Delta,\lambda)d\Delta d\lambda = \bar{P}d\Delta d\lambda$ where the distribution $\bar{P}$ is a constant. This assumption leads to a density of states $n(E)$ which is approximately constant and which results in good agreement with acoustic and dielectric measurements. Problems, however, arise in the interpretation of fast specific heat measurements as discussed in Chap.3.

As already mentioned the description given so far completely fails in the interpretation of the nonsaturable part of the absorption in the glassy metal PdSi [6.57]. Obviously, the electrons also interact with the two-level systems resulting in extremely short values of T_1 [6.79a]. Therefore, the assumption $\omega T_1 \gg 1$ leading to (6.20) for the relaxation absorption does not hold even at temperatures as low as 30 mK. The absorption increases with temperature although $\omega T_1 \ll 1$. Nevertheless, the experimental data can be fitted if one assumes that the density of tunneling states is not constant, but increases for higher values of λ [6.57].

The microscopic nature of the tunneling particle is still an open question. So far it is not clear whether the same type of defect is responsible for both absorption peaks described in this chapter. It is likely that single atoms — for example, an oxygen atom in the case of vitreous silica — are responsible for the thermally activated relaxation process observed at nitrogen temperatures. Measurements of the dielectric absorption suggest that the particle tunneling at low temperatures is more complex; it is possible to attach an OH-group to the tunneling particle and thus to change its dipole moment without drastically changing the other properties of the two-level system. In order to learn more about the atomic nature of the two-level systems, further experimental and theoretical studies are desirable where concrete models for the moving particle are investigated.

6.7 Summary

The acoustic and dielectric properties of amorphous materials at low temperatures are fundamentally different from those of their crystalline counterparts. In glasses the acoustic and dielectric absorption is qualitatively very similar. A more or less pronounced peak, or in some cases a broad shoulder, is found around the temperature of liquid nitrogen. A second absorption maximum is observed at helium temperatures in many amorphous materials. At still lower temperatures, below 1 K, the absorption rises again, but can be "saturated" at higher intensities. A characteristic temperature dependence of the velocity of sound and the dielectric constant is associated with these absorption processes. Quantitatively there is an important difference between the acoustic and dielectric behavior at low temperatures. Whereas the acoustic properties are only modestly influenced by impurities, the magnitude of the variation of the complex dielectric constant is mainly determined by the content of polar impurities.

Only two different mechanisms are needed to explain the complex behavior of glasses. The attenuation at higher temperatures is thought to be thermally activated and is attributed to particles—probably single atoms—moving in double-well potentials with a broad distribution of barrier heights. At lower temperatures the attenuation is caused by the interaction between phonons and two-level systems with a broad distribution of energy splitting. Both acoustic and dielectric measurements can be used to study the dynamics of the low-energy excitations observed in amorphous materials at low temperatures. Their description in terms of two-level systems is based on these experiments and has turned out to be extremely successful. Dielectric measurements also give information on the possible microscopic nature of these systems. Very recently similar experiments have also been carried out in disordered crystalline materials [6.94]. Qualitatively identical results have been obtained, indicating that the occurrence of acoustic and dielectric anomalies at low temperatures is not restricted to amorphous materials but can be observed in a wide variety of substances.

Acknowledgements. We want to thank Prof. K. Dransfeld and Drs. W. Arnold, J.L. Black, H. Sussner, and R. Vacher for stimulating discussions and helpful comments during the preparation of this manuscript.

References

6.1   J.V. Fitzgerald: J. Am. Ceram. Soc. *34*, 314-319 (1951)
6.2   H.J. McSkimin: J. Appl. Phys. *24*, 988-997 (1953)
6.3   M.E. Fine, H. Van Duyne, N.T. Kenney: J. Appl. Phys. *25*, 402-405 (1954)
6.4   O.L. Anderson, H.E. Bömmel: J. Am. Ceram. Soc. *38*, 125-131 (1955)
6.5   J.T. Krause: J. Am. Ceram. Soc. *47*, 103 (1964)
6.6   J.T. Krause: Phys. Lett. *43*A, 325-326 (1973)
6.7   R.E. Strakna, H.T. Savage: J. Appl. Phys. *35*, 1445-1450 (1964)
6.8   C.K. Jones, P.G. Klemens, J.A. Rayne: Phys. Lett. *8*, 31-32 (1964)
6.9   A.S. Pine: Phys. Rev. *185*, 1187-1193 (1969)
6.10 C. Krischer: J. Acoust. Soc. Am. *48*, 1086-1092 (1970)
6.11 D.B. Fraser, J.T. Krause, A.H. Meitzler: Appl. Phys. Lett. *11*, 308-310 (1967)
6.12 J.T. Krause, C.R. Kurkjian: J. Am. Ceram. Soc. *51*, 226-227 (1968)
6.13 C.R. Kurkjian, J.T. Krause: J. Am. Ceram. Soc. *49*, 134-138 (1966)
6.14 C.A. Maynell, G.A. Saunders, S. Scholes: J. Non Cryst. Solids *12*, 271-294 (1973)
6.15 J. Jäckle, L. Piche, W. Arnold, S. Hunklinger: J. Non Cryst. Solids *20*, 365-391 (1976)
6.16 R. Vacher, J. Pelous: Phys. Rev. B*14*, 823-826 (1976)

104

6.17 M. Dutoit: Phys. Lett. *50*A, 221-223 (1974)
6.18 B.S. Berry, W.C. Pritchet, C.C. Tsuei: Phys. Rev. Lett. *41*, 410-413 (1978)
6.19 J.T. Krause, C.R. Kurkjian, D.A. Pinnow, E.A. Sigety: Appl. Phys. Lett. *17*, 367-368 (1970)
6.20 H. Kaga, S. Kashida, S. Umehara: J. Phys. Soc. Jpn. *44*, 1208-1215 (1978)
6.21 J.T. Krause: J. Appl. Phys. *42*, 3035-3037 (1971)
6.22 J.M. Stevels: "The Electrical Properties of Glass", in *Electrical Conductivity II*, Encyclopedia of Physics, Vol.XX, ed. by S. Flügge (Berlin, Heidelberg, New York, Springer 1967) pp.350-391
6.23 R.E. Jaeger: J. Am. Ceram. Soc. *51*, 57-58 (1968)
6.24 E.M. Amrhein: Glastech. Ber. *42*, 52-57 (1969)
6.25 S.M. Mahle, R.D. McCammon: Phys. Chem. Glasses *10*, 222-225 (1969)
6.26 W.W. Scott, R.K. McCrone: Phys. Rev. B*1*, 3515-3524 (1970)
6.27 M. Mansingh, J.M. Reyes, M. Sayer: J. Non Cryst. Solids *7*, 12-22 (1972)
6.28 M.G.R. Zobel: Thesis, University of Bristol (1970)
6.29 U. Strom, P.C. Taylor: Phys. Rev. B*16*, 5512-5522 (1977)
6.30 S. Hunklinger: Ultrasonic Symposium Proceedings (IEEE, New York 1974) pp.493-501
6.31 R.E. Strakna: Phys. Rev. *123*, 2020-2026 (1961)
6.32 M.R. Vukcevich: J. Non Cryst. Solids *11*, 25-63 (1972)
6.33 P.W. Anderson: Phys. Rev. Lett. *34*, 953-955 (1975)
6.34 M. Kastner, D. Adler, H. Fritzsche: Phys. Rev. Lett. *37*, 1504-1507 (1976)
6.35 N.F. Mott: Adv. Phys. *26*, 363-391 (1977)
6.36 G.N. Greaves: Philos. Mag. *37*, 447 (1978)
6.37 M.N. Kulbitskaya, S.V. Nemilov, V.A. Shutilov: Sov. Phys. Solid State *16*, 2319-2322 (1975)
6.38 R. Vacher, J. Pelous: Phys. Lett. *53*A, 233-235 (1975)
6.39 M.D. Mashkovich, E.N. Smelyanskaya: Sov. Phys. Solid State *9*, 974-975 (1967)
6.40 R.D. McCammon, R.N. Work: Rev. Sci. Instrum. *36*, 1169-1173 (1965)
6.41 M.N. Kulbitskaya, V.A. Shutilov: Sov. Phys. Acoust. *22*, 451-461 (1976)
6.42 N. Thomas: Thesis, University of Cambridge (1978)
6.42a M. Von Haumeder, U. Strom, S. Hunklinger: Phys. Rev. Lett. *44*, 84 (1980)
6.42b M. Von Haumeder: Thesis, University of Konstanz (1980)
6.43 S. Hunklinger, W. Arnold, S. Stein: Phys. Lett. A*45*, 311-312 (1973)
6.44 P. Doussineau, A. Levelut, G. Bellessa, O. Bethoux: J. Phys. Paris Lett. *38*, L 483-487 (1977)
6.45 G. Bellessa, M. Cagnon, J.F. Sadoc, P. Doussineau, A. Levelut: J. Phys. Paris Lett. *37*, L 291-293 (1976)
6.45a G. Weiss, W. Arnold, K. Dransfeld, H.J. Güntherodt: Solid State Commun. *33*, 111 (1980)
6.46 G. Frossati, J. Le G. Gilchrist, J.C. Lasjaunias, W. Meyer: J. Phys. C*10*, L 515-519 (1977)
6.47 W. Heinicke, G. Winterling, K. Dransfeld: J. Acoust. Soc. Am. *49*, 954-958 (1971)
6.48 S. Hunklinger, W. Arnold, S. Stein, R. Nava, K. Dransfeld: Phys. Lett. A*42*, 253-255 (1972)
6.49 B. Golding, J.E. Graebner, B.I. Halperin, R.J. Schutz: Phys. Rev. Lett. *30*, 223-226 (1973)
6.50 B. Golding, J.E. Graebner, R.J. Schutz: Phys. Rev. B*14*, 1660-1662 (1976)
6.51 W. Arnold, S. Hunklinger, S. Stein, K. Dransfeld: J. Non Cryst. Solids *14*, 192-200 (1974)
6.52 A. Bachellerie: In *Satellite Symposium of the Eigth International Congress on Acoustics on Microwave Acoustics*, ed. by E.R. Dobbs, J.K. Wigmore (Institute of Physics, London 1974) pp.93-96
6.53 S. Hunklinger: Festkörperprobleme (Adv. Solid State Phys.) *17*, 1-11 (1977)
6.54 M. Von Schickfus, S. Hunklinger: Phys. Lett. *64*A, 144-146 (1977)
6.55 M. Von Schickfus: Thesis, University of Konstanz (1977)
6.56 P. Doussineau, P. Legros, A. Levelut, A. Robin: J. Phys. Paris Lett. *39*, L 265-269 (1978)

6.57   B. Golding, J.E. Graebner, A.B. Kane, J.L. Black: Phys. Rev. Lett. *41*, 1487-1490 (1978)
6.58   L. Pichê, R. Maynard, S. Hunklinger, J. Jaeckle: Phys. Rev. Lett. *32*, 1426-1429 (1974)
6.59   S. Hunklinger, L. Pichê: Solid State Commun. *17*, 1189-1192 (1975)
6.60   J. Pelous, R. Vacher: Solid State Commun. *19*, 627-630 (1976)
6.61   J. Pelous, R. Vacher: J. Phys. Paris Lett. *38*, 1153-1159 (1977)
6.62   B. Golding, J.E. Graebner, A.B. Kane: Phys. Rev. Lett. *37*, 1248-1250 (1976)
6.63   S. Hunklinger, W. Arnold: In *Physical Acoustics*, ed. by W.P. Mason, R.N. Thurston (Academic, New York 1976), Vol.XII, pp.155-215
6.64   C. Laermans, L. Piche, W. Arnold, S. Hunklinger: In *Non-Crystalline Solids*, Proc. 4th Int. Conf., ed. by G.H. Frischat (TransTech, Aedermannsdorf 1977) pp.562-567
6.65   G. Bellessa, C. Lemercier, D. Caldemaison: Phys. Lett. *62*A, 127-128 (1977)
6.66   G. Bellessa: J. Phys. C*10*, L 285-287 (1977)
6.67   G. Bellessa, P. Doussineau, A. Levelut: J. Phys. Paris Lett. *38*, L 65-66 (1977)
6.68   J.M. Farley, G.A. Saunders: J. Non Cryst. Solids *18*, 417-427 (1975)
6.69   M. Von Schickfus, S. Hunklinger, L. Pichê: Phys. Rev. Lett. *35*, 876-878 (1975)
6.70   M. Von Schickfus, C. Laermans, W. Arnold, S. Hunklinger: In *Non-Crystalline Solids*, Proc. 4th Int. Conf., ed. by G.H. Frischat (TransTech, Aedermanns-dorf 1977) pp.542-547
6.71   M. Von Schickfus, S. Hunklinger: J. Phys. C*9*, 439-442 (1976)
6.72   B. Golding, J.E. Graebner, W.H. Haemmerle: In *Amorphous and Liquid Semi-conductors*, ed. by W.E. Spear (CTCL University of Edinburgh 1977) pp.367-371
6.73   C. Laermans, W. Arnold, S. Hunklinger: J. Phys. C*10*, L 161-165 (1977)
6.74   P. Doussineau, A. Levelut, Ta Thu-Thuy: J. Phys. Paris Lett. *38*, L 37-39 (1977)
6.75   P. Doussineau, A. Levelut, Ta Thu-Thuy: J. Phys. Paris Lett. *39*, L 55-57 (1978)
6.76   R.C. Zeller, R.O. Pohl: Phys. Rev. B *4*, 2029-2041 (1971)
6.77   R.B. Stephens: Phys. Rev. B*8*, 2896-2905 (1973)
6.78   J.C. Lasjaunias, A. Ravex, M. Vandorpe, S. Hunklinger: Solid State Commun. *17*, 1045-1049 (1975)
6.79   J.A. Sussmann: J. Phys. Chem. Solids *28*, 1643-1648 (1967)
6.79a   J.L. Black, P. Fulde: Phys. Rev. Lett. *43*, 453 (1979)
6.80   J. Joffrin, A. Levelut: J. Phys. Paris Lett. *36*, 811-822 (1975)
6.81   W. Arnold, S. Hunklinger: Solid State Commun. *17*, 883-886 (1975)
6.82   J.L. Black, B.I. Halperin: Phys. Rev. B*16*, 2879-2895 (1978)
6.83   W. Arnold, C. Martinon, S. Hunklinger: J. Phys. Paris Lett. *39*, C-6, 961-962 (1978)
6.84   A. Abragam: In *The Principle of Nuclear Magnetism*, ed. by N.F. Mott, E.C. Bullard, D.H. Wilkinson (Oxford University Press, London 1961)
6.85   J.D. Macomber: *The Dynamics of Spectroscopic Transitions* (Wiley, New York 1976)
6.86   J. Jaeckle: Z. Phys. *257*, 212-223 (1972)
6.87   S.L. McCall, E.L. Hahn: Phys. Rev.*183*, 457-485 (1969)
6.88   P.J. Davis: In *Handbook of Mathematical Functions*, ed. by T.A. Stegun, M.A. Abramovitz (Dover, New York 1964)
6.89   W. Arnold, J.L. Black, G. Weiss: In *Phonon Scattering in Condensed Matter*, ed. by H.J. Maris (Plenum, New York 1980) pp.77-80
6.90   S. Hunklinger, L. Piche, J.C. Lasjaunias, K. Dransfeld: J. Phys. C*8*, L423-426 (1975)
6.91   P.W. Anderson, B.I. Halperin, C. Varma: Philos. Mag. *25*, 1-9 (1972)
6.92   W.A. Phillips: J. Low Temp. Physics *7*, 351-360 (1972)
6.93   V. Narayanamurti, R.O. Pohl: Rev. Mod. Phys. *42*, 201-236 (1970)
6.94   U. Strom, M. Von Schickfus, S. Hunklinger: Phys. Rev. Lett. *41*, 910-913 (1978)

7. Relaxation Times of Tunneling Systems in Glasses

B. Golding and J. E. Graebner

With 10 Figures

The purpose of this chapter is 1) to review experiments which provide information on two-level system relaxation times in glasses, 2) to relate these relaxation times to the specific physical processes responsible for them, and 3) to use this information to expand our knowledge of two-level systems, their interactions with each other and external fields. Most of the experiments reported by early 1979 have been carried out on various types of silica glasses, but relaxation time information is becoming available on systems such as metallic glasses.

   The plan of this chapter is the following: First, the resonance dynamics of two-level systems is reviewed and extended to tunneling levels in glass. Phenomenological relaxation times are introduced into equations of motion of observable quantities. Next, physical processes believed responsible for T_1 and T_2 in various glasses are described. Then, experiments which provide information on the magnitudes and the temperature and energy dependence of these times are reviewed. These experiments are arbitrarily organized by those falling into incoherent or coherent regimes, and by 1, 2, or 3 pulse experimental methods. Finally, we discuss to what extend the physical models and experiment are consistent with one another.

7.1 Background

Glasses possess physical properties strikingly different from crystalline solids at low temperatures ($T \cong 1$ K). This fact has been appreciated, however, for considerably less than a decade. It was not until 1971, when measurements of the thermal and transport properties of insulating glasses were extended below 1 K [7.1], that it became clear that amorphous solids apparently possess extra low energy excitations not present in crystalline media. The observation in these experiments of a large quasi-linear contribution to the specific heat and of unusually small thermal phonon mean free paths (inferred from thermal conductivities) led to theoretical suggestions [7.2,3] that glasses consist of large numbers of tunneling systems which interact strongly with phonons. The central ideas of the tunneling model are: 1) the existence of a wide distribution of the sizes of local interatomic or intermolecular potentials due to the structural disorder; 2) the possibility of

atoms or molecules undergoing quantum-mechanical tunneling between double-well
potentials at low temperatures; 3) that only the lowest two energy levels of each
tunneling entity need be considered; and 4) the presence of a strong deformation-
potential coupling between these energy levels and long-wavelength strain fields
(phonons). Within this model a glass is viewed as an ensemble of two-level systems
whose energy splittings are described by a continuous distribution function P(E),
finite from essentially E = 0 to energies ~kT_g with T_g the glass transition temperature. The specific heat, nearly linear in T below 1 K, then arises from the
thermal excitation of two-level systems possessing nearly constant P(E) for
$E/k_B \lesssim 1$ K. The small thermal conductivity (phonon mean free path) occurs as a result of strong resonant absorption of thermal phonons by the two-level systems.

Although specific heats and thermal conductivities provide necessary constraints
on any theoretical model of the low-temperature properties of glasses, they involve
averages over energies on the order of $k_B T$ and, thus, are not sensitive to details
on a finer energy scale. On the other hand, the study of the interaction of nearly
monochromatic radiation (acoustic, electromagnetic) with the two-level systems is
capable of revealing information about the two-level systems in an arbitrarily fine
spectral region. The earliest example was a form of acoustic spectroscopy in which
nonlinear sound propagation at phonon frequencies near 1 GHz was observed in silica
glass below 1 K [7.4,5]. This immediately confirmed the idea of resonant absorption
of phonons by systems possessing only two energy levels since at high acoustic intensities the absorption saturated, i.e., was driven to zero by equalizing the populations of the two levels. It was shown [7.5] that the formalism developed to describe magnetic resonance in a spin-1/2 system [7.6] or two-level microwave or optical resonance [7.7] could be applied to advantage with virtually no modification
in interpreting phonon resonance in glasses.

The interpretation of resonance phenomena, and indeed the existence of resonant
absorption itself, depends upon the exchange of energy between the resonant two-
level systems and some thermal reservoir. Otherwise, at any level of excitation
the populations of the two energy levels would eventually equalize and no steady-
state absorption could be realized. In glasses, as in other resonant systems, thermal equilibrium is brought about by the emission and absorption of phonons (or by
electron interactions). The rate at which equilibration takes place is denoted by
T_1^{-1}, and T_1 is the analog of the "spin-lattice" relaxation time of magnetic resonance [7.6].

In considering resonant absorption by the two-level systems it is inappropriate
to visualize the two levels as being infinitely sharp in energy. In reality, the
levels are broadened or have a finite energy width; this phenomenon occurs for
two distinct reasons. One form of broadening, *inhomogeneous broadening*, results
from a static distribution of resonance energies due to the different environment
of each two-level system. This is, of course, an intrinsic feature of a glass and

results in an inhomogeneous spectral width T_2^{*-1} determined by the extent of $P(E)$, i.e., T_2^{*-1} is effectively infinite. In actual experiments the spectral range of systems that are probed will be governed by the radiation spectrum. Another and more fundamental type of broadening for glasses, *homogeneous broadening*, can be understood by considering only a single two-level system. Here, a level will have a finite energy width due to decay (T_1 process) and the uncertainty principle, or as a result of other processes which may cause the energy splitting to vary, such as interactions with other perturbations, e.g., other fields or two-level systems. It is customary to define a homogeneous relaxation rate $T_2'^{-1} \approx T_1^{-1} + T_\phi^{-1}$ where T_ϕ is the phase memory time which arises from processes in which there may be no net exchange of energy between a two-level system and another system.

These relaxation times play a central role in understanding the classes of resonance phenomena which are observable in any given system. If one considers the excitation of a resonant system by a pulse of resonant radiation with duration τ, then the classes of phenomena can be conveniently grouped in the following way:

$T_1, T_2' \ll \tau$. This is the steady-state or cw regime since initial transients decay rapidly compared to the total excitation time.

$T_2' < \tau < T_1$. In this regime energy decays from an excited two-level system in a time comparable to or longer than the excitation pulse. The transient behavior now assumes importance since a steady state is never reached. This regime, like the preceding one, is characterized as incoherent since phase memory is not maintained throughout the excitation period.

$\tau \ll T_1, T_2'$. This is termed the coherent regime since there is no loss of energy or phase memory during excitation.

It is a remarkable fact that all of these regimes have been observed in acoustic resonance in one type of glass, fused silica, by varying the temperature between 0.02 and 1.5 K.

7.2 Resonance Dynamics of Two-Level Systems

We shall consider here the behavior of a general two-level quantum mechanical system exposed to a resonant radiation field, treated classically, which is capable of exciting transitions between its energy states. We use a standard density matrix formalism to describe a system coupled to a harmonic field. The field may be of acoustic, electric, or magnetic origin, although we shall emphasize those interactions for which nonvanishing two-level system phonon matrix elements exist in describing glasses. The density-matrix method has found extensive use not only in magnetic resonance literature [7.8] but, more recently, in describing optical resonance [7.9,10], and we apply it here anticipating that more sophisticated resonance phenomena remain yet to be discovered in glasses. This general treatment of resonance is then applied specifically to the atomic tunneling model of glasses [7.2,3]. This involves the introduction of energy eigenvalues and interactions which are a function of tunneling parameters. The tunneling parameters do not generally possess unique values but are broadly distributed. The physical origin of the re-laxation processes in glasses is then described. The longitudinal relaxation time T_1 in insulating glasses is assumed to arise from a one-phonon (direct) process [7.2,3,11], and the phase memory time T_ϕ is taken to originate from a spectral dif-fusion process [7.12] due to interactions of two-level systems with each other. In metallic glasses conduction electrons provide an alternate decay path [7.13,14].

A two-level system is described by the model Hamiltonian

$$H = H_0 + H_1 + H_2 = \frac{1}{2} E\sigma_z - \left(M\sigma_x + \frac{1}{2} D\sigma_z\right)e(t) - \left(\mu'\sigma_x + \frac{1}{2} \mu\sigma_z\right)F(t) \quad . \tag{7.1}$$

The first term is the static Hamiltonian with eigenvalues $\pm \frac{1}{2} E$ whereas the second and third terms represent, respectively, the interaction of the system with the time-dependent strain field $e(t)$ and electric field $F(t)$. Transitions between the two levels are effected via the general matrix elements Me and $\mu'F$. The quantities M and D are, respectively, the off-diagonal and diagonal deformation potentials, while μ' and μ are, respectively, the induced and static electric dipole moments. The terms in σ_z lead to shifts in E and are important in describing dephasing pro-cesses due to interacting two-level systems as well as nonresonant absorption and dispersion. The σ's are the Pauli spin operators

$$\sigma_x = \begin{pmatrix} 0 & 1 \\ 1 & 0 \end{pmatrix} \quad \sigma_y = \begin{pmatrix} 0 & -i \\ i & 0 \end{pmatrix} \quad \sigma_z = \begin{pmatrix} 1 & 0 \\ 0 & -1 \end{pmatrix} \quad . \tag{7.2}$$

The system wave function is written as

$$\Psi(r,t) = C_a(t)u_a(r) + C_b(t)u_b(r) \tag{7.3}$$

where subscripts a and b refer, respectively, to the upper $(+ \frac{1}{2} E)$ and lower $(- \frac{1}{2} E)$ energy levels. The observables of this system, which are obtained by cal-culating expectation values with respect to $\Psi(r,t)$, involve combinations of the time-varying amplitudes $C_a(t)$ and $C_b(t)$ [7.10]. We write the density matrix for an N-particle system with all particles possessing the same Hamiltonian as

$$\rho = \begin{pmatrix} \rho_{aa} & \rho_{ab} \\ \rho_{ba} & \rho_{bb} \end{pmatrix} = \sum_j^N P_j \begin{pmatrix} C_{aj}C_{aj}^* & C_{aj}C_{bj}^* \\ C_{bj}C_{aj}^* & C_{bj}C_{bj}^* \end{pmatrix} \tag{7.4}$$

where P_j is the fraction of systems with wave function $\Psi_j(r,t)$. The diagonal elements ρ_{ii} represent the probability that a particular two-level system is in level i. In terms of the density matrix the expectation value of an operator <A> is given by

$$<A> = NTr\{\rho A\} \qquad (7.5)$$

and for normalized wave functions one has

$$Tr\{\rho\} = 1 \quad . \qquad (7.6)$$

The density matrix obeys the equation of motion

$$\dot{\rho} = -i\hbar[H,\rho] \qquad (7.7)$$

in which we have momentarily neglected damping terms. If we consider, for example, a time-varying excitation of the form

$$e(t) = e_0 \exp(-i\omega t) \qquad (7.8)$$

in the rotating-wave approximation, we obtain for the off-diagonal matrix elements of the strain interaction Hamiltonian

$$(H_1)_{ab} = -(1/2)Me_0 e^{-i\omega t} \quad . \qquad (7.9)$$

It is customary to calculate the time dependence of the density matrix in an interaction picture in which the rapid time dependence at ω is neglected, equivalent to a transformation to a coordinate system rotating at ω [7.8,10]. We define

$$\rho' = \begin{pmatrix} \rho_{aa} & \overline{\rho_{ab}}\exp(i\omega t) \\ \overline{\rho_{ba}}\exp(-i\omega t) & \rho_{bb} \end{pmatrix} \qquad (7.10)$$

and solve (7.7) for $\dot{\rho}'$. We also define in terms of the elements of the density matrix the components of a pseudopolarization vector $\underline{P} = Tr(\rho'\sigma) = u\hat{u} + v\hat{v} + w\hat{w}$,

$$u = \rho'_{ab} + \rho'_{ba} \qquad (7.11)$$

$$v = i(\rho'_{ab} - \rho'_{ba}) \qquad (7.12)$$

$$w = \rho'_{aa} - \rho'_{bb} \quad . \qquad (7.13)$$

The equation of motion obeyed by $\underline{P}$ is

$$\dot{\underline{P}} = \underline{\omega} \times \underline{P} \qquad (7.14)$$

where $\underline{\omega} \equiv (-Me_0/\hbar, 0, \omega - \omega_0)$ and $E = \hbar\omega_0$. The individual components of $\dot{P}$ are

$$\dot{u} = \frac{-u}{T'_2} - (\omega - \omega_0)v \qquad (7.15)$$

$$\dot{v} = (\omega - \omega_0)u - \frac{v}{T'_2} + \left(\frac{Me_0}{\hbar}\right)w \qquad (7.16)$$

$$\dot{w} = \frac{-Me_0 v}{\hbar} - \frac{(w - w_0)}{T_1} \qquad (7.17)$$

in which we have introduced phenomenological damping times T_1 and T_2' to describe, respectively, the decay of the diagonal and off-diagonal elements of the density matrix. In addition, electric field driving terms will also exist provided $\mu' \neq 0$. In the absence of excitation the population difference w relaxes to its thermal equilibrium value $w_0 = -\tanh(E/2k_BT)$. The above equations, known as the Bloch equations [7.6,8], were first derived for the case of nuclear magnetic resonance and are generalized here for any two-level system. Equations (7.15-17) show that for on-resonance excitation ($\omega - \omega_0 = 0$), and neglecting damping, a population can be inverted by application of a pulse of amplitude e_0 and duration τ such that

$$\theta \equiv \omega_1\tau = \frac{Me_0\tau}{\hbar} \tag{7.18}$$

is equal to π. It is customary to refer to θ as the pulse area. Thus, rotations of arbitrary angles of the polarization vector may be effected by the application of resonant or near-resonant radiation pulses. This is pictured in Fig.7.1b and 7.1c in which the system, initially in its ground state at $t = 0$, i.e., $w = w_0 = -1$ is found in a state $w = 0$, $v = -1$ after application of a $\theta = \pi/2$ pulse at ω_0.

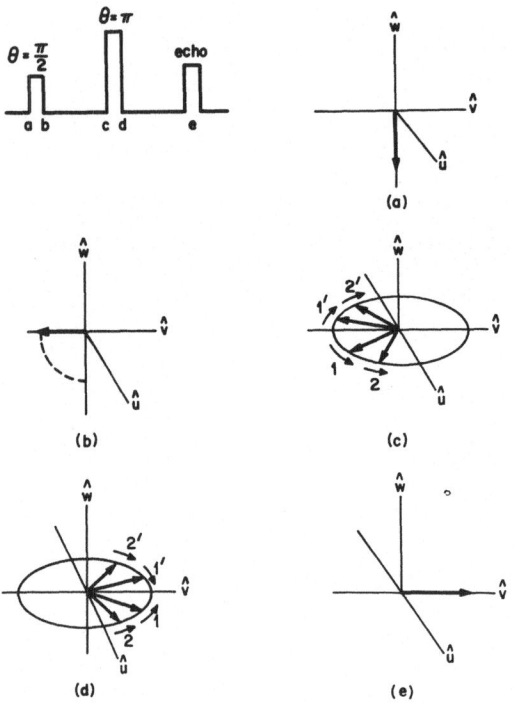

Fig.7.1. Pulse sequence and polarization vector rotations in a rotating frame describing the generation of a two-pulse echo in an inhomogeneously broadened system

7.2.1 Tunneling Model

The tunneling model [7.2,3] provides a rationale for the existence of intrinsic two-level systems in glasses. It is assumed that the static potential fluctuations due to structural disorder may be decomposed into nearly equivalent double potential wells separated by a barrier of height V between which atoms or groups of atoms may undergo quantum-mechanical tunneling. The coupling of the two wells is described by the quantity $\Delta_0 = \hbar\Omega e^{-\lambda}$, in which $\hbar\Omega$ is a typical zero-point energy of the particle of mass m in a well, and the overlap parameter $\lambda = \hbar^{-1}d(2mV)^{1/2}$ for harmonic wells separated by coordinate d. The energy eigenvalues of this model are given by

$$E = \pm \frac{1}{2}(\Delta^2 + \Delta_0^2)^{1/2} \qquad (7.19)$$

where the asymmetry Δ is the difference in the potential energy minima of the two wells, with $\Delta = 0$ for symmetric wells. The coupling parameters in the basis which diagonalizes the two-well Hamiltonian to yield (7.19) give interaction terms which are functions of tunneling parameters

$$M = \frac{\Delta_0}{E}\gamma \qquad (7.20)$$

and

$$D = 2\frac{\Delta}{E}\gamma \qquad (7.21)$$

where $\gamma \equiv (1/2)(\partial\Delta/\partial e)$ and e is the appropriate perturbing field. When e is a strain tensor γ will also be a tensor quantity, and appropriate averages over its components will need to be taken since the orientation of the two-level system will vary from site to site [7.12,15]. The coupling brought about by terms of the form $(\partial\Delta_0/\partial e)$ is neglected since it is several orders of magnitude weaker than coupling through the asymmetry Δ [7.12,16]. It is generally assumed that a probability distribution $\bar{P}(\lambda)$ exists, independent of Δ, for two-level systems under consideration here, which gives the number of two-level systems per increment $d\Delta$ and $d\lambda$. The quantity λ varies from its minimum value $\lambda_{min}(E) = \ln(\hbar\Omega/E)$ to an undetermined maximum λ_{max}, and for $\lambda_{min} < \lambda < \lambda_{max}$ it is usually assumed that $\bar{P}(\lambda) = \bar{P}$, a constant. Since the off-diagonal coupling to external fields is a maximum when $\lambda = \lambda_{min}$, for which $M = \gamma$, resonant processes are dominated by those two-level systems possessing $\lambda \approx \lambda_{min}$, which may be a fraction of those two-level systems contributing to the specific heat [7.5,17]. A phonon can be resonantly absorbed when its energy $\hbar\omega$ coincides with the energy splitting of the levels of a two-level system, $\hbar\omega = E$. This leads to a frequency- and temperature-dependent reciprocal decay length, or attenuation coefficient [7.2,3,5],

$$l_\alpha^{-1} = \frac{\pi\omega}{\rho v_\alpha^3}\bar{P}\gamma_\alpha^2 \tanh\left(\frac{\hbar\omega}{2k_BT}\right), \qquad (7.22)$$

where v_α is the phase velocity of a phonon of polarization α (longitudinal or transverse, L or T) and ρ is the mass density of the glass. From a variety of ex-periments, it has been found that for silica and silica-based glasses, generally, $\gamma_L^2 \approx 2\gamma_T^2$ [7.18,19].

7.2.2 Longitudinal Relaxation Time T_1

In the absence of an external driving field the diagonal elements of the density matrix relax to time independent equilibrium values. Let $W_{ab}(W_{ba})$ be the transition probability from state a to b (b to a) when e = 0. Thus, we obtain the rate equa-tions [7.9]

$$\dot{\rho}_{aa} = \rho_{bb}W_{ba} - \rho_{aa}W_{ab} \qquad (7.23)$$

$$\dot{\rho}_{bb} = \rho_{aa}W_{ab} - \rho_{bb}W_{ba} \quad . \qquad (7.24)$$

In thermal equilibrium each ρ_{ii} should be time independent leading to the relation

$$\rho_{bb}^0 W_{ba} = \rho_{aa}^0 W_{ab} \quad . \qquad (7.25)$$

We now define the longitudinal relaxation time T_1 as

$$T_1 \equiv \frac{\rho_{bb}^0}{W_{ab}} = \frac{\rho_{aa}^0}{W_{ba}} \quad . \qquad (7.26)$$

Note that this is equivalent to defining

$$T_1^{-1} = W_{ab} + W_{ba} \quad . \qquad (7.27)$$

Thus

$$\dot{\rho}_{aa} = T_1^{-1}[\rho_{bb}\rho_{aa}^0 - \rho_{aa}\rho_{bb}^0] \qquad (7.28)$$

$$\dot{\rho}_{bb} = T_1^{-1}[\rho_{aa}\rho_{bb}^0 - \rho_{bb}\rho_{aa}^0] \qquad (7.29)$$

and since $\text{Tr}\{\rho\} = 1$ we find

$$\dot{\rho}_{ii} = -T_1^{-1}[\rho_{ii} - \rho_{ii}^0] \qquad (7.30)$$

$$\dot{w} = \dot{\rho}_{aa} - \dot{\rho}_{bb} = -T_1^{-1}(w - w^0) \qquad (7.31)$$

showing that the individual diagonal elements of ρ and the inversion, w, relax ex-ponentially to their thermal equilibrium value. In thermal equilibrium $\rho_{ii} = \exp(-E_i/kT)/\sum_i \exp(-E_i/kT)$ so that $w^0 = -\tanh(E/2k_BT)$, $E = E_i - E_j$.

a) *One-Phonon Relaxation*

The transition probabilities for a one-phonon process are calculated by the golden rule

$$W_{ab} = \sum_{\alpha} \frac{2\pi}{\hbar} |<b|H_1|a>|^2 g(E)[1 + n_{ph}(E)]\delta(\hbar\omega - E) \qquad (7.32)$$

$$W_{ba} = \sum_{\alpha} \frac{2\pi}{\hbar} |<a|H_1|b>|^2 g(E)n_{ph}(E)\delta(\hbar\omega - E) \qquad (7.33)$$

where we define $n_{ph}(E) = \{\exp(E/k_B T - 1\}^{-1}$ and the summation is over phonon polarizations α. The matrix elements [7.11] are given by

$$<b|H_1|a> = <a|H_1|b> = M_{\alpha}(\hbar/2\rho\omega)^{\frac{1}{2}}k_{\alpha} \qquad (7.34)$$

and the phonon density of states in the range $E/k_B \lesssim 1$ K is

$$g(E) = E^2/2\pi^2\hbar^3 v_{\alpha}^3 \quad . \qquad (7.35)$$

Therefore, one finds generally

$$T_1^{-1}(E) = \sum_{\alpha} \left(\frac{M_{\alpha}^2}{v_{\alpha}^5}\right) \frac{E^3}{2\pi\rho\hbar^4} \coth\left(\frac{E}{2k_B T}\right) \qquad (7.36)$$

and within the tunneling model using (7.20)

$$T_1^{-1}(E) = \sum_{\alpha} \left(\frac{\gamma_{\alpha}^2}{v_{\alpha}^5}\right)\left(\frac{E\Delta_0^2}{2\pi\rho\hbar^4}\right)\coth\left(\frac{E}{2k_B T}\right) \quad . \qquad (7.37)$$

The dependence of Δ_0 on λ can be expressed by the relation $\Delta_0 = Ee^{-(\lambda-\lambda min)}$ so that the final result of the calculation is

$$T_1^{-1}(E,\lambda) = \left[\left(\frac{\gamma_L^2}{v_L^5}\right)+\left(\frac{2\gamma_T^2}{v_T^5}\right)\right]E^3 \frac{e^{-2(\lambda-\lambda min)}}{2\pi\rho\hbar^4} \coth\left(\frac{E}{2k_B T}\right) \quad . \qquad (7.38)$$

The energy dependence of T_1^{-1} will vary between E^3, when $x = E/2k_B T \gtrsim 1$, and E^2, when $x \ll 1$ and $\coth(x) \approx \frac{1}{x}$. The major contribution to relaxation arises from emission and absorption of transverse phonons, by virtue of their greater state density. If a glass consists of a reasonably large range of λ's, it is clear that an overall *nonexponential* relaxation of w can be expected with a maximum rate for those two-level systems with $\lambda = \lambda_{min}$. At temperatures well above 1 K more effective higher-order phonon processes will undoubtedly dominate two-level relaxation rates [7.20]. Detailed studies of resonant processes in this region have not been made due to the appearance of strong relaxational absorption and will not be considered further here.

b) *Conduction Electron Relaxation*

Tunneling systems exist in metallic glasses with densities of states similar to insulating glasses such as a-SiO$_2$ [7.21,22,23]. However, the dynamics of the tunneling systems can be quite different at temperatures well below 1 K where T_1 may be 4 orders of magnitude shorter than in insulators. At these temperatures phonon processes cannot explain such large rates, and a coupling of the tunneling levels to the conduction electrons has been invoked to explain this phenomenon [7.13,14]. An interaction Hamiltonian can be written as

$$H_3 = 1/2(v_\perp \sigma_x + v_{\shortparallel} \sigma_z) \sum_{kk'} c^+_{k'} c_k \tag{7.39}$$

where the c_k's are electron operators, and $v_\perp$ and $v_{\shortparallel}$ are, respectively, the off-diagonal and diagonal parameters coupling conduction electrons to tunneling centers. The electrons will renormalize the basic tunneling energies, i.e., will contribute to H_0, but this effect will not be discussed here [7.24,25].

The specific relaxation process which has been considered is the inelastic scattering of an electron from state k to k', differing in energy by $E = \varepsilon_{k'} - \varepsilon_k$.

The transition probabilities are [7.8]

$$W_{ab} = \frac{2\pi}{\hbar} \sum_{kk'} |<b|H_3|a>|^2 f_k (1 - f_{k'}) \delta(\varepsilon_k - \varepsilon_{k'} + E) \tag{7.40}$$

$$W_{ba} = \frac{2\pi}{\hbar} \sum_{kk'} |<a|H_3|b>|^2 f_k (1 - f_{k'}) \delta(\varepsilon_k - \varepsilon_{k'} - E) \tag{7.41}$$

where f_k is the Fermi function

$$f_k = f(\varepsilon_k) = \{\exp[(\varepsilon_k - \varepsilon_F)/k_B T] + 1\}^{-1} . \tag{7.42}$$

We write

$$W_{ab} = \frac{\pi}{2\hbar} (v_\perp^2) \int_0^\infty d\varepsilon_k d\varepsilon_{k'} n(\varepsilon_k) n(\varepsilon_{k'}) f_k (1 - f_{k'}) \delta(\varepsilon_k - \varepsilon_{k'} + E) \tag{7.43}$$

$$= \frac{\pi}{2\hbar} (v_\perp^2) \int_0^\infty d\varepsilon_k n^2(\varepsilon_k) f_k [1 - f(\varepsilon_k + E)] , \tag{7.44}$$

where $n(\varepsilon)$ is the single spin density of states. Carrying out the integral and recognizing that the major contribution comes from electrons at the Fermi level with energy ε_F we find

$$W_{ab} = \frac{\pi}{4\hbar} [N(\varepsilon_F) v_\perp]^2 E [1 - \exp(-E/k_B T)]^{-1} . \tag{7.45}$$

Invoking (7.27) we obtain the result

$$T_1^{-1}(E) = \frac{\pi}{4\hbar} [N(\varepsilon_F) v_\perp]^2 E \coth(E/2k_B T) \tag{7.46}$$

where $N(\varepsilon_F)$ is the total electronic density of states at the Fermi level. This result differs by a factor of 2 from Reference [7.14] in which the expression for T_1^{-1} contained the density of states for a particular spin orientation.

In the high-temperature limit, expansion of $\coth(x) \approx 1/x$ yields the energy independent expression $T_1 T = \text{constant}$. This bears a close resemblance to the Korringa relaxation of nuclear spins in metals by the conduction electrons, and, in fact, the processes are formally analogous [7.26,8].

The major difference between the electronic and the single phonon relaxation processes lies in the much weaker E-dependence of the former process. As has been described elsewhere [7.14], even though electronic relaxation may dominate relaxation of thermally excited tunneling systems in the mK range, at temperatures near 1 K the one-phonon process may take over due to its E^3 dependence.

7.2.3 Transverse Relaxation Time T_2'

The relaxation of the off-diagonal components of the density matrix is given by the homogeneous tranverse relaxation time T_2', see (7.15,16). T_2^{-1} may be viewed as the intrinsic spectral width of a homogeneous group of two-level systems centered on ω_0. Two known sources of homogeneous broadening exist in glasses: those which arise from longitudinal relaxation, and those which arise from processes affecting the phase of the precessing two-level systems with, at most, a minor shift in their energy (compared to E). Where this separation is clear one has

$$T_2'^{-1} = \frac{1}{2} T_1^{-1} + T_\phi^{-1} \tag{7.47}$$

in which a phase memory time T_ϕ is defined. In glasses the origin of T_ϕ is interactions of the two-level systems with each other. One process well known in magnetic resonance [7.8] is the mututal flipping of systems of the same energy by emission of a resonant phonon. In glasses, however, the density of resonant two-level systems lying within a typical pulse spectrum is only $\sim 10^{12}$ cm^{-3}, too dilute for this process to dominate dephasing. A more effective dephasing process involves the interaction of all thermally activated two-level systems with the resonant species via a σ_z term of (7.1). This process may exist since $D\sigma_z$ is an elastic dipole moment and acts as a stress source (driving term) when nonresonant two-level systems undergo a change of state, or "flip". Since two-level systems with energy $\lesssim 2k_B T$ are continually undergoing excitation and decay by thermal phonon interactions, a resonant two-level system may be subjected to a random time-varying elastic stress which causes its energy splitting to fluctuate. The interaction between spins is described by the addition of a term to the Hamiltonian [7.12,16]

$$H_4 = \frac{1}{4} \sum_{i<j} J_{ij} \sigma_z^i \sigma_z^j \tag{7.48}$$

in which

$$J_{ij} = c_{ij}(\Delta_i/E_i)(\Delta_j/E_j)r_{ij}^{-3} \quad , \tag{7.49}$$

where r_{ij} is the distance between two-level systems at site i and site j, and c_{ij} depends on deformation-potential couplings. Because of the r_{ij}^{-3} dependence this elastic dipolar interaction is formally similar to magnetic or electric dipole couplings.

In the case of glasses the line broadening originating in (7.48) may, in fact, be a time-dependent quantity describable in terms of a process known as spectral diffusion [7.12,27,28]. A spectral diffusion kernel $D(\omega - \omega_0, t)$ is defined as the probability that a two-level system of initial energy $E = \hbar\omega_0$ at $t = 0$ has energy $\hbar\omega$ at time t,

$$D(\omega - \omega_0, t) = \frac{1}{\pi} \frac{\Delta\omega(t)}{(\omega - \omega_0)^2 + [\Delta\omega(t)]^2} \quad . \tag{7.50}$$

In the limit of short times following excitation of two-level systems at $\hbar\omega_0$, i.e., $t \ll T_1(E \approx 2k_BT)$, an explicit expression for $\Delta\omega(t)$ has been given [7.12] as

$$\Delta\omega(t,T) = \Delta\omega(\infty,T) \frac{t}{2(\lambda_{max} - \lambda_{min})} \int_0^\infty T_1^{-1}(x)\mathrm{sech}^2 x \, dx \quad . \tag{7.51}$$

From (7.38) we have $T_1^{-1} = AT^3 x^3 \coth(x)$, with $A = (\gamma_L^2/v_L^5 + 2\gamma_T^2/v_T^5)4k_B^3/\pi\rho\hbar^4$, and the integration yields

$$\Delta\omega(t,T) = \Delta\omega(\infty,T) \frac{A}{2(\lambda_{max} - \lambda_{min})} \frac{\pi^4}{64} T^3 t \tag{7.52}$$

where

$$\Delta\omega(\infty,T) = \frac{\pi^2}{3\hbar} (<c_{ij}^2>)^{\frac{1}{2}} (\lambda_{max} - \lambda_{min})\bar{P} k_B T \quad . \tag{7.53}$$

The above equations indicate a spectral width $\Delta\omega$ which varies linearly with time (at short times) but which approaches a limiting value $\Delta\omega(\infty)$ at long times. For small t the spectral width is strongly temperature dependent, varying as T^4. Calculations have been performed which show the time evolution of the width for arbitrary times, using realistic parameters for silica glass [7.12].

7.2.4 Spontaneous Echo Decay

The observation of phonon and electric echoes in silica glass [7.29,30,31] has led to a direct and accurate determination of T_2' when this time is comparable to, or larger than, experimentally attainable excitation pulse widths. The echo appears as a third pulse of resonant radiation [7.28,32,33] occuring after the resonant systems are prepared by an appropriate two-pulse sequence, as depicted in Fig.7.1. With the ensemble initially in thermal equilibrium ($w = w^0$, $u = v = 0$), a pulse of area

$\theta_1 = \pi/2$ puts the two-level system into a nonstationary state, Fig.7.1(b), which has associated with it a maximum induced transverse polarization. This polarization decays to zero Fig.7.1(c) as a result of the inhomogeneous nature of the excited system. The band of excited systems contains some two-level systems with resonant frequencies larger and smaller than ω_0 and these precess about w at differing rates, thus rapidly dephasing. The second pulse, with $\theta_2 = \pi$, rotates each two-level system about the û axis, Fig.7.1(d), and since the sense of precession remains unchanged, at a time equal to that between θ_1 and θ_2, all two-level systems are rephased momentarily along -v̂, Fig.7.1(e). Again, a net polarization is developed and a coherent acoustic or electric pulse is emitted. In reality, not all two-level systems thus excited will contribute to the echo, since there may be phase-disrupting processes affecting a certain fraction of the two-level systems between the time of the initial pulse and the appearance of the echo, a time interval $2\tau_{12}$. In this simple picture one expects the echo amplitude to decay as $E(2\tau_{12}) = E(0)\exp(-2\tau_{12}/T_2')$. Thus, a measurement of E as a function of $2\tau_{12}$ leads to a determination of the homogeneous relaxation time, even in the presence of large inhomogeneous broadening.

As stated above, spectral diffusion is believed to be the major contributor to the loss of phase memory in glasses since the shift of energy of a resonant two-level system due to its interaction with other two-level systems leads to a loss of phase coherence. For a two-pulse sequence the echo amplitude is given by [7.28, 12]

$$E(2\tau_{12}) = <\exp[i \int_0^{\tau_{12}} \omega(t)dt - i \int_{\tau_{12}}^{2\tau_{12}} \omega(t)dt]> \qquad (7.54)$$

where <...> denotes an average over all resonant two-level systems and over all flip histories of the thermally excited nonresonant two-level systems. Again, in the short-time limit the spectral diffusion kernel yields the result [7.12,27,28]

$$E(2\tau_{12}) = E(0)\exp(-m\tau_{12}^2) \qquad (7.55)$$

in which $m = t^{-1}\Delta\omega(t,T)$ as $t \to 0$. Although in this limit echo decay is nonexponential, a "time constant" T_ϕ may be defined by setting $mT_\phi^2 = 1$, so that $T_\phi = m^{-\frac{1}{2}} \sim T^{-2}$, using (7.52) for $\Delta\omega(t,T)$. Situations in which the short-time limit does not apply have been investigated [7.34] and in these regimes more complex decay patterns are predicted.

7.2.5 Stimulated Echo Decay

If, after the two-pulse sequence described above, a third pulse of resonant radiation is applied to the glass at a time τ_{13}, additional echoes may be created. For $\tau_{13} > 2\tau_{12}$ the echo which immediately follows the third pulse at $\tau_{13} + \tau_{12}$ is known as a stimulated echo [7.32]. Unlike the spontaneous echo, the stimulated echo does not depend on the persistence of phase memory until $\tau_{13} + \tau_{12}$. Instead, the requirement is that the populations of the two-level systems, put into a non-equilibrium state by the first two pulses, have not reached thermal equilibration at the echo time. The stimulated echo is, therefore, dependent on T_1 processes.

In the standard description [7.32,35] the excitation consists of three $\pi/2$ pulses. The first pulse creates the nonstationary situation depicted in Fig.7.1(b). Dephasing occurs as in Fig.7.1(c) but the second $\pi/2$ rotation reorients the dipoles to lie in the $\hat{u}-\hat{w}$ plane. Since precession about the $\hat{w}$ axis still exists, each dipole traces out a cone whose principal axis is $\hat{w}$ with cone angle ϕ representing the amount of dephasing at time (c). During the interval between τ_{12} and τ_{13}, each cone relaxes back to the ground state along $-\hat{w}$ according to $\exp(-t/T_1)$. At τ_{13}, the third $\pi/2$ pulse reorients the cone axis along $\pm v$, and a rephasing of cone edges occurs at $\tau_{13} + \tau_{12}$ creating a finite polarization, the stimulated echo. In this simple picture the stimulated echo amplitude E_{st} decays as

$$E_{st}(\tau_{13},\tau_{12}) = \exp[-(\tau_{13} + \tau_{12})/T_1] \quad . \tag{7.56}$$

when $\tau_{12} << T_2'$.

Spectral diffusion produces enhanced decay and deviations from this exponential form [7.12,28]. Its effect can be seen from the fact that the first two closely spaced pulses produce a frequency comb with the pulse spectrum modulated at τ_{12}^{-1}. This comb can decay either by a T_1 process or by spectral diffusion of the excited systems over a fraction of τ_{12}^{-1}. In the usual pulsing conditions this fraction is a small excursion in frequency compared to τ^{-1}. Thus, the decay of the stimulated echo may be appreciably enhanced over (7.56) if rapid spectral diffusion is present.

In a way similar to (7.54) one defines the spectral diffusion contribution to stimulated echo decay

$$E_{sd}(\tau_{13},\tau_{12}) = \left\langle \exp\left[i \int_0^{\tau_{12}} dt\omega(t) - i \int_{\tau_{13}}^{\tau_{13}+\tau_{12}} dt\omega(t) \right] \right\rangle . \tag{7.57}$$

In the short time limit, $\tau_{12},\tau_{13} << T_1 (E \approx 2k_BT)$ it has been found that [7.12,28],

$$E_{sd}(\tau_{13},\tau_{12}) = \exp(-m\tau_{12}\tau_{13}) = \exp(-m\tau_{12}^2)\exp[-\overline{\tau_{12}m}(\tau_{13} - \tau_{12})]. \tag{7.58}$$

Therefore, even if $\tau_{13} >> \tau_{12}$, which is the usual situation in glass echo experiments, τ_{12} can continue to significantly influence the rate of echo decay. This parametric dependence on τ_{12} of the stimulated echo decay is perhaps the most direct test of the presence of spectral diffusion [7.12,28].

7.3 Experiments Measuring Relaxation Times

The experiments to be described cannot be classified conveniently according to the relaxation time measured since some experiments measure a *combination* of T_1 and T_2'. All measurements described here involve transient methods, so that experiments are ordered, rather arbitrarily, by the number of excitation pulses utilized.

Recent experiments in pure and OH-doped fused silica have revealed the presence of more than one type of tunneling system [7.36]. When comparing experimental results, therefore, it is important to keep in mind differences in sample material. Most of the measurements with insulating glasses discussed here were performed on three types of commercially available glass: Suprasil W (SiO_2 with < 2 ppm OH), Suprasil I (SiO_2 with 1200 ppm OH), and BK7 (a borosilicate glass).

7.3.1 Acoustic Saturation

The first prediction of the two-level tunneling model to be verified experimentally [7.4,5] was the nonlinear acoustic attenuation which occurs when the acoustic intensity is strong enough to modify significantly the populations of the two energy levels. The crossover from high absorption at low input intensity to low absorption at high intensity locates the critical intensity, I_c. At I_c, the rate of excitation ω_1 is balanced against the decay rates T_1^{-1} and $(T_2')^{-1}$. The steady state solution to the Bloch equations (7.15-17) yields [7.5,16,37] $\alpha = \alpha_0 (1 + I/I_c)^{-\frac{1}{2}}$, where α_0 is the unsaturated absorption and $I_c = \hbar^2 \rho v^3 / 2\gamma^2 T_1 T_2'$. As will be discussed below, many of the measurements of I_c have used pulse lengths shorter than T_1, so that steady state conditions did not apply.

7.3.2 Saturation Recovery

In a two-pulse saturation recovery experiment the first pulse is strong enough to saturate the resonant two-level systems and the attenuation of the weak second pulse at a time τ_{12} later is used to monitor the recovery of the population difference. If the spectrum of excited states does not change between the two pulses, the experiment measures the time for equilibration of the two-level systems with the thermal reservoir, i.e., T_1. The first results [7.5] were obtained in Suprasil I at 0.4 K with 0.5 μs pulses at 2.0 GHz with $\tau_{12} \geq 0.2$ μs, and established an upper limit of ~0.2 μs on T_1. The second measurement of saturation recovery [7.37] used 1 μs pulses at 760 MHz with $0.3 < T < 0.8$ K. Results were similar for both BK7 and vitreous silica. To convert the absorption change $\Delta\alpha$, which is the experimentally measured quantity, to the population difference an expression for $\Delta\alpha$ derived from the steady state solutions to the Bloch equations was used [7.37]. A recovery time of ≈ 0.8 μs at 0.5 K with a roughly T^{-1} dependence over the temperature range of the experiment was observed.

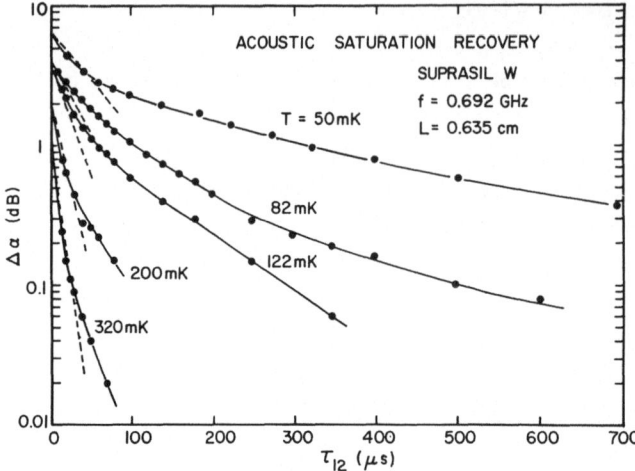

Fig.7.2. The change in acoustic attenuation of a weak test pulse P_2 at time τ_{12} after a strong saturating pulse P_1 of the same frequency [7.39]. The initial values of $\Delta\alpha$ were obtained from saturation curves. The pulse intensities, which were not the same for all temperatures, lay in the range $2 - 8 \times 10^{-6}$ W/cm^2 for P_1 and $0.2 - 2 \times 10^{-7}$ W/cm^2 for P_2

Saturation recovery data in Suprasil W are also available at lower temperatures [7.38,39] over the range of τ_{12} shown in Fig.7.2. The results were obtained in the transient regime ($T_2' < \tau < T_1$) using a probing-pulse intensity which was small enough to assure no worse than 10% nonlinearity between $\Delta\alpha$ and population differ-ence [7.39]. The recovery to equilibrium is seen to be very temperature dependent and to extend out to several milliseconds at 50 mK.

By contrast, saturation recovery measurements in a *metallic* glass [7.14] have indicated an extremely short T_1. The recovery at 0.96 GHz and 10 mK in Pd$_{.775}$Si$_{.165}$ Cu$_{.06}$ was considerably shorter than the smallest pulse duration used, about 100 ns. An upper limit on T_1 was placed at 25 ns, which is four orders of magnitude shorter than the decay time in insulating glasses under comparable conditions. This is due to the conduction electrons which relax the two-level systems, as discussed in Sect.7.2.2b.

7.3.3 Linewidth

Because of the large inhomogeneous broadening in glasses it is possible to sa-turate only a small fraction of the total linewidth. The resulting hole burned into the spectrum can be probed in a second type of two-pulse experiment. If the time separation of the two pulses in a saturation recovery experiment remains fixed but the frequency difference is varied, the attenuation of the probing pulse as a function of the frequency difference traces out the spectral hole created by the saturating pulse. The first experiment of this kind [7.40] used 1 μs pulses

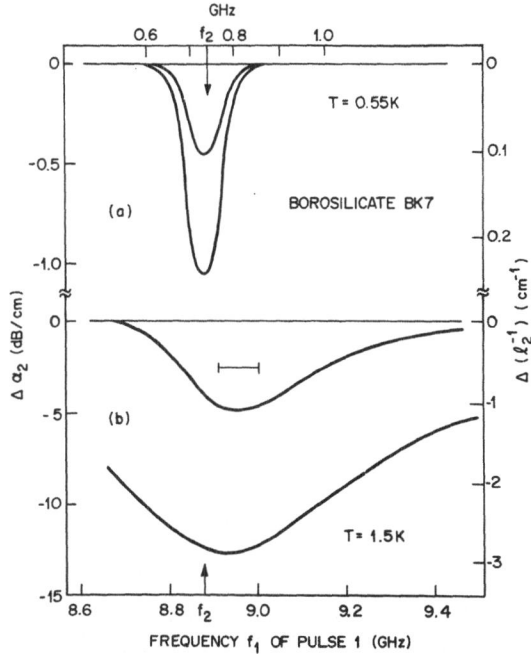

Fig.7.3. The change in acoustic attenuation of a test pulse at frequency f_2 in two holeburning experiments, (a) [7.37] and (b) [7.41], for two values of the saturating pulse (P_1) intensity in each experiment. The intensities were 5×10^{-6} and 5×10^{-7} W/cm^2 for P_1 and 0.9×10^{-7} W/cm^2 for P_2 in (a). Absolute intensities were not measured in (b) but the two values of intensity for P_1 differed by a factor of 33. The bracket in (b) indicates the width (fwhm) deduced for the low-power limit, i.e., in the absence of power broadening

propagating simultaneously in BK7 glass. The data are shown in Fig.7.3a with two different intensities for the saturating pulse at 0.55 K. The linewidth (half-width at half-maximum) is 45 MHz at 0.55 K and 65 MHz at 0.79 K, i.e., the width is proportional to T. Note that the linewidth is more than an order of magnitude greater than the pulse excitation spectrum. Furthermore, the linewidth is independent of the intensity of the saturating pulse for the range of intensity used.

A second linewidth experiment [7.41] in BK7 near 9 GHz and 1.5 K again utilized simultaneous 1 μs pulses. Here, the linewidth depended on the power in either pulse (Fig.7.3b) varying at high intensities as the square root of the intensity. This is consistent with the concepts of power broadening as derived from steady state solutions of the Bloch equations, where the effective linewidth contains [7.6] a factor $(1 + I/I_c)^{\frac{1}{2}}$. The analysis of the data required a value of the unsaturated attenuation, which was not measured. Assuming a value for the unsaturated attenuation extrapolated from lower frequency data, a value of 45 MHz was calculated for the linewidth in the absence of power broadening.

Another linewidth experiment [7.39] used 0.50 μs pulses at 0.69 GHz with separation time between leading edges of 1 μs in Suprasil W in the temperature range 80-200 mK. Below ~100 mK, the width of the spectral hole was accounted for by the spectrum of the pulse and was assumed to be temperature independent [7.42]. At 122 mK, however, an additional full width of 1 MHz was observed and at 200 mK, 8 MHz.

A fourth linewidth experiment [7.43] used simultaneous pulses with variable pulse length in Suprasil I in an effort to measure the *time-dependent* linewidth occurring in spectral diffusion (Sect.7.2.3). At T = 0.42 K and with a center frequency of 0.78 GHz, the linewidth increased from 14 to 24 MHz as the pulse length was increased from 0.3 to 1.3 μs.

Because the linewidth becomes narrower than the spectrum of a conveniently long pulse at low temperatures, the above method is limited to measuring linewidths above ~100 mK. At lower temperatures different techniques must be used.

7.3.4 Two-Pulse Phonon Echo

At low enough temperatures the phase memory time is longer than the pulse width and coherent effects such as the phonon echoes described in Sect.7.2.4 become observable. The spontaneous phonon echo [7.29,31] was generated by two identical input pulses which travelled through the sample and were reflected from the opposite face. The reflected pulses were observed as R_{11} and R_{21} (Fig.7.4), followed by the phonon echo pulse E_{12}^1 at a time τ_{12} after R_{21}. At each point along the acoustic path the input pulses caused the coherent dephasing and rephasing described in Sect.7.2.4, resulting in an acoustic pulse propagating behind the second input pulse. The pulse labelled E_{12}^2 is a second echo which was formed by the second input pulse and the first echo acting as generating pulses. The phase memory time was measured by observing the change in amplitude of E_{12}^1 as τ_{12} was varied, as shown in Fig.7.5 for four different temperatures. The decay was approximately exponential with a time constant which varied rapidly with temperature. At somewhat higher or lower input amplitudes the decay for small τ_{12} was generally not exponential [7.31] but at longer τ_{12} appeared to approach the decay rates shown in Fig.7.5. The time constants at intermediate amplitude, Fig.7.6, vary as T^{-2} between 18 and 80 mK with a decay time of 16 μs at 18 mK.

7.3.5 Two-Pulse Electric Echo

The spontaneous electric echo was first observed [7.30] by applying 0.93 GHz electric pulses to a capacitor with dielectric material consisting of a 0.5 μm film of sputtered SiO_2 containing no known dipolar impurities. Echo decay times of 10 and 8 μs (Fig.7.7) were obtained at 15 and 20 mK, respectively, similar to phonon echoes at 0.69 GHz.

Spontaneous electric echoes have also been observed in *bulk* samples of fused silica, with OH impurities [7.36,44-47] as well as without OH [7.36]. It has been shown [7.36] that there are two distinct tunneling systems in a-SiO_2, one associated with the OH impurities ($\mu' = 3.7$ D) and the other associated with the intrinsic tunneling systems of the glass ($\mu' = 0.6$ D). By appropriate selection of sample material and electric field amplitude, either species could be made to dominate the echo response. At 19 mK and 0.7 GHz the echo generated from the intrinsic (int)

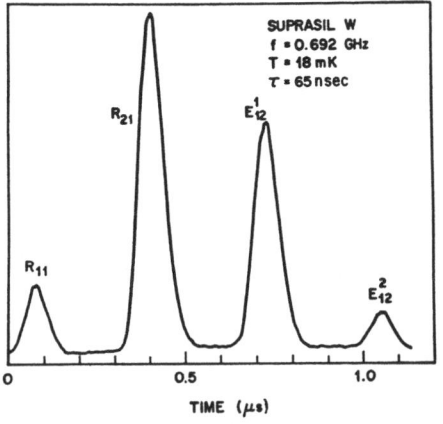

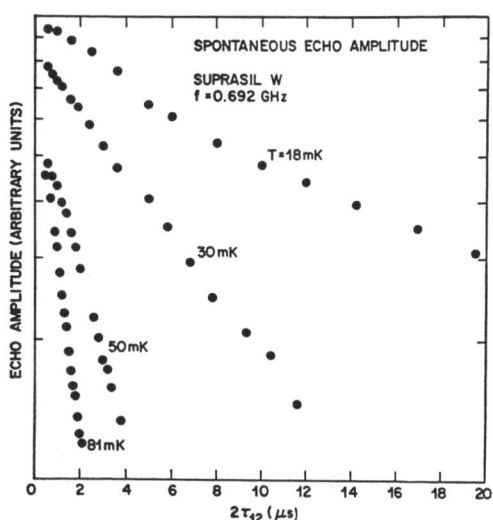

Fig.7.4. Spontaneous phonon echo [7.31]. R_{11} and R_{12} are the input pulses observed after one round trip of the sample, while E_{12}^1 is the spontaneous phonon echo. E_{12}^2 is a secondary echo generated by R_{21} and E_{12}^1. Pulse width of the identical input pulses is 65 ns, pulse separation $\tau_{12} = 0.33$ μs

Fig.7.5. Decay of the spontaneous phonon echo amplitude [7.31]. Input pulses are identical with $\tau = 65$ ns and energy density 8.1×10^{-7} erg/cm^2 per pulse

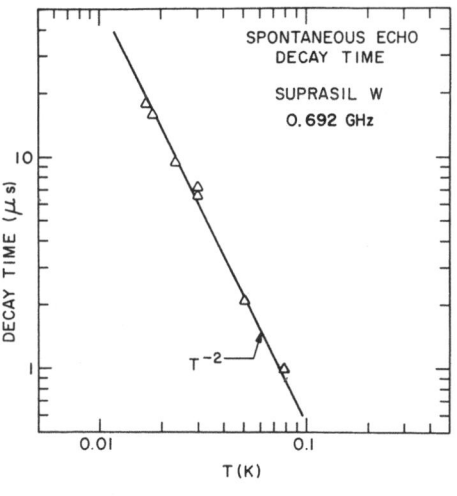

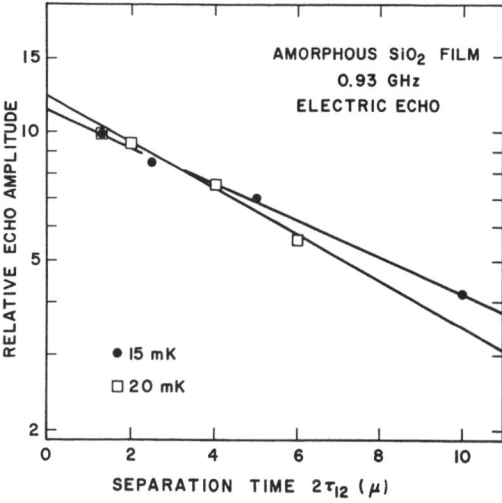

Fig.7.6. Decay time of the spontaneous phonon echo vs temperature [7.31]. The level of the input pulses was the same for all the data, at 8.1×10^{-7} erg/cm^2 per pulse

Fig.7.7. Decay of the spontaneous electric echo [7.39]

species decayed nearly exponentially, with $T_2'(int) = 16$ µs, in excellent agreement with phonon echo measurements in OH-free fused silica. The echo decay of the OH-associated species yielded $T_2'(OH) \approx 25$ µs.

An analysis of earlier measurements at 0.37 GHz on an OH-containing Suprasil I sample [7.46,47] suggested that T_2' varied as T^{-1} between 4 and 20 mK, as opposed to the T^{-2} behavior of spontaneous phonon echoes at 0.69 GHz above 18 mK.

7.3.6 Three-Pulse Phonon Echo

The stimulated phonon echo is generated with the three-pulse sequence described in Sect.7.2.5 and has been observed in Suprasil W in the same range of frequency and temperature as the spontaneous phonon echo. The echo amplitude is shown in Fig.7.8 as a function of the delay time of the third pulse. At all temperatures investigated the echo amplitude decreased rapidly at first but much more slowly at large τ_{13}. The shape of the decay curve did not depend on the amplitude of the input pulses as for the spontaneous echo. Because the decays in Fig.7.8 are not exponential, it is not possible to describe the entire relaxation by a single time constant. If the *initial* decay time is plotted vs T, a roughly T^{-2} dependence is observed with a time constant of ~100 µs at 20 mK, for $\tau_{12} = 0.5$ µs. The shape of the stimulated echo decay depends on τ_{12} [7.31]. As τ_{12} is increased the overall decay becomes faster, especially at small τ_{13}.

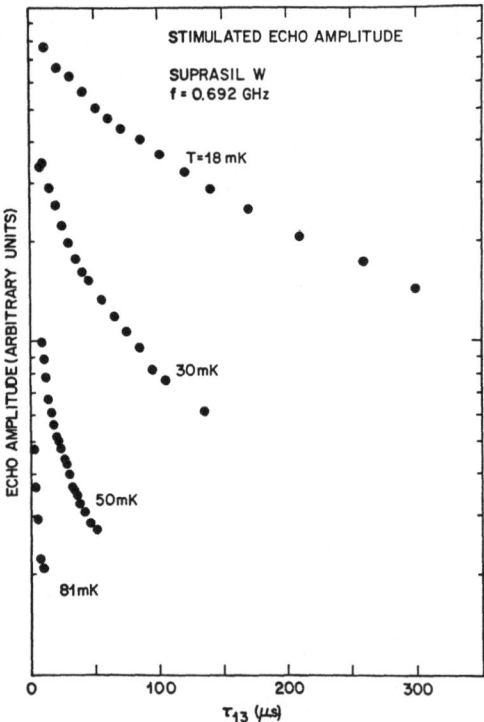

STIMULATED ECHO AMPLITUDE

SUPRASIL W
f = 0.692 GHz

T=18 mK

30mK

50mK

81mK

ECHO AMPLITUDE (ARBITRARY UNITS)

τ_{13} (µs)

Fig.7.8. Decay of the stimulated phonon echo [7.31]. Input pulses are identical with $\tau = 65$ ns

7.3.7 Three-Pulse Electric Echo

The stimulated electric echo observed in the thin film of SiO_2 [7.30] exhibited an initial decay time of 50 µs at 15 mK and 0.93 GHz.

The stimulated electric echo observed in bulk silica containing OH [7.46] exhibited a nonexponential decay similar to that of the stimulated phonon echo, but with an initial decay time of ~100 µs at 20 mK and 0.37 GHz which varied approximately as $\tanh(hf/2k_BT)$ to 4 mK. No dependence on τ_{12} was noted.

The more recent measurements [7.36] on both OH and OH-free glasses at 19 mK have revealed distinctly different recovery times for the intrinsic and OH-associated tunneling species. The single-double pulse sequence utilized [7.6], is inherently less sensitive to spectral diffusion than the stimulated echo sequence discussed above. The insensitivity exists because the appearance of the echo does not depend on the persistence of a spectral comb. In this technique the first pulse inverts the two-level systems. After a time τ_{12}, which may be much greater than T_2', the second and third pulses at fixed separation from each other generate a spontaneous echo whose amplitude is proportional to the longitudinal polarization at time τ_{12}. Recovery times of 140 and 410 µs were observed for the intrinsic and OH-related two-level systems, respectively, at 19 mK and 0.72 GHz.

7.4 Critical Assessment of Data

7.4.1 T_1 Results

Results of the two very different types of experiments in insulators involving T_1, i.e., three-pulse echoes and saturation recovery, yield relaxation times which lie generally in the range appropriate to the one-phonon process. The magnitude and temperature dependence of the apparent T_1, however, suggest that additional decay mechanisms may be simultaneously operative.

a) *Three-Pulse Echoes*

The nonexponential decay of the stimulated echo has been noted above. One mechanism of decay, spectral diffusion, should affect the echo according to (7.58). Calculations [7.12] based on (7.58) yield envelopes which resemble the data of Fig.7.8. Quantitative comparison is difficult but the initial decay time can be estimated from the calculations. This time is 100 µs at 20 mK and varies approximately as T^{-2}, in agreement with the data [7.31]. More significantly, the initial decay rate depends on the separation time of the first two pulses in a way which is understandable in terms of spectral diffusion (Sect.7.2.5).

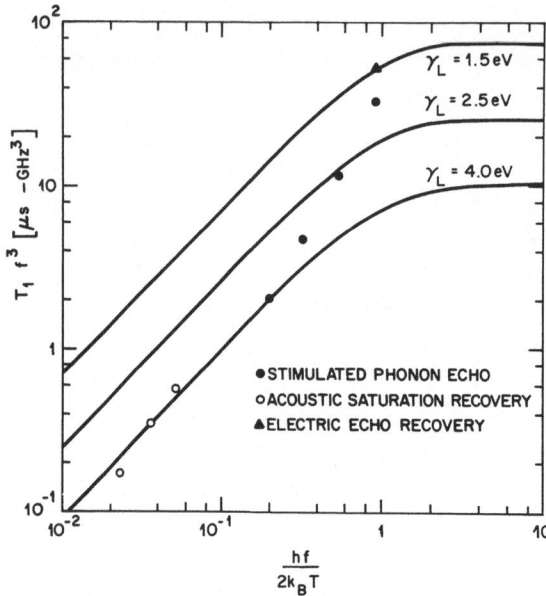

Fig.7.9. Comparison of the apparent recovery time with T_1 calculated for the one-phonon process (7.38) for three values of the deformation potential, using material parameters for Suprasil W. Data are included for the stimulated phonon echo [7.31], the one-two sequence electric echo [7.36], and acoustic saturation recovery at 0.76 GHz [7.37]

Another, more intrinsic, factor which may affect the stimulated echo decay is the distribution of decay times expected within the tunneling model (Sect.7.2.2a). A small distribution in λ results in a large range of decay times, according to (7.38). Thus, the apparent spread in decay times of the data in Fig.7.8 may be partially explained by the distribution of λ.

The initial decay times of the stimulated phonon echoes are plotted in Fig.7.9 as $T_1 f^3$ vs $x = hf/2k_BT$ to facilitate comparison of results at different frequencies with the predictions of the direct process, (7.38). The data are approximated by a T^{-2} temperature dependence, as opposed to the T^{-1} or weaker behavior expected with a simple direct process, and are in qualitative agreement with the effects of spectral diffusion [7.12] referred to above. It seems likely, then, that the lifetime of the stimulated phonon echo is determined by spectral diffusion which obscures the slower direct process (with $\gamma_L = 1.5$ eV), even at the lowest temperature measured.

The three-pulse, one-two sequence electric echo time constant, 140 μs, for the intrinsic tunneling levels at 0.72 GHz is also plotted in Fig.7.9. It agrees very well with the one-phonon prediction for $\gamma_L = 1.5$ eV.

b) *Saturation Recovery*

The simple excited spectrum of a saturation recovery experiment is less susceptible to distortion by spectral diffusion than the modulated spectrum of the stimulated echo experiment. Thus, saturation recovery is a potentially more accurate measure of the return to equilibrium. The saturation recovery data of Fig.7.2 exhibit very long tails similar to the stimulated echo decays, but the initial decay times are generally longer than for stimulated echoes. The decay times based on a one-phonon process with γ_L = 1.5 eV are plotted as dashed lines in Fig.7.2. The similarity to the initial decays is apparent. We note, again, the possibility that a distribution of tunneling parameters may be responsible for the typically slow decay at long times.

Saturation recovery data in vitreous silica at higher temperatures [7.37] are represented by the three points at small x in Fig.7.9. Note that the decay times calculated from the data are comparable to the pulse width, so that the assumption of steady state conditions is only marginally valid. Perhaps a more serious problem is the *apparent* recovery of the center of the line due to the time-dependent line broadening observed under similar conditions [7.43]. That is, the hole burned into the spectrum becomes broader as well as becoming more shallow. If the line broadening is indeed caused by spectral diffusion, one would expect it to affect more significantly the data [7.37] taken with a relatively narrow pulse spectrum (τ = 1 μs, T > 0.3 K) than the saturation recovery data [7.39] at lower temperatures (τ = 0.06 μs, T < 0.3 K) with a broader initial spectrum and slower spectral diffusion. This argument is qualitatively consistent with the difference between the saturation recovery data in Fig.7.2 and that in Fig.7.9.

Considerable care must therefore be taken when interpreting data related to thermal equilibration. In both the stimulated phonon echo and the high-temperature long-pulse saturation recovery experiments, spectral diffusion masks the one-phonon recovery of the system. The low-temperature short-pulse saturation recovery and the one-two sequence electric echo experiments, however, are relatively unaffected by spectral diffusion and lend strong support to one-phonon relaxation with γ_L = 1.5 eV.

c) *Distribution of Decay Times*

Evidence for a distribution of decay times has been noted, particularly in the long tails of the saturation recovery data (Fig.7.2). Further evidence comes from acoustic measurements in a metallic glass [7.14] and dielectric measurements in an insulating glass [7.48,49].

The extremely short T_1 of the metallic glass is attributed (Sect.7.2.2b) to relaxation of the two-level systems by the conduction electrons, and allows the regime ωT_1 < 1 to be explored. The effect of the fast T_1 process on the relaxational (nonresonant) acoustic absorption and dispersion has been calculated within the formalism of the tunneling model and fitted to both absorption and dispersion [7.14] over a temperature range 0.01 - 10 K. The parameters in the calculation were also constrained by the resonant absorption and dispersion processes. A broad spectrum of decay rates was required to fit the data, since no absorption maximum as a function of T was observed, as would be expected for a unique T_1. Two-level systems with energy splittings $E \approx k_B \times 0.1$ K, for example, were found to require

relaxation times distributed over a range of at least four orders of magnitude
$(\Delta\lambda = \lambda_{max} - \lambda_{min} \geq 5)$.

An analysis of the temperature- and frequency-dependent dielectric constant of oxide glasses has been pursued along somewhat similar lines [7.48,49]. In one case [7.48] a range of relaxation times of four orders of magnitude was used to fit the data.

While it is difficult to deduce exact values for $\Delta\lambda$, the above results provide strong evidence for a distribution of relaxation times for two-level systems of the same energy, an idea which is central to the two-level tunneling model.

7.4.2 T_2' Results

a) *Two-Pulse Echoes*

The decay of the spontaneous echo is the most direct measurement of T_2' that is available at low temperatures (below 100 mK). Though there are uncertainties due to the amplitude dependence of the shape of the decays noted above, the decay times in Fig.7.6 are described well by a T^{-2} behavior with $T_2' = 16$ μs at 18 mK for intrinsic tunneling systems in Suprasil W (at temperatures above 18 mK). Comparison can be made between the data (Fig.7.5 and Fig.7.6) and theoretical predictions based on spectral diffusion. The echo decay envelopes in Fig.7.5 do not agree with the $\exp(-m\tau_{12}^2)$ behavior of (7.55) calculated in the short-time limit. On the other hand, the line proportional to T^{-2} in Fig.7.6 is the time constant T_ϕ calculated on the basis of spectral diffusion [7.12] (see the discussion following (7.55). The exact agreement may be fortuitous, considering the approximately 30% uncertainty in the value of γ_L used in the calculation. The agreement of both magnitude *and* temperature dependence, however, gives support to the picture of spectral diffusion. The disagreement with the decay shape may have to do with the pulses traveling back and forth over the same acoustic path several times, causing multiple excitation of the resonant species. The theory does not take into account the propagation effects present in the acoustic experiment. This cannot, however, be the full explanation because the decay of the (nonpropagating) electric echoes is genérally very similar to that of the phonon echoes. The failure of (7.55) to describe the decay may also have to do with the restriction to the short-time limit [7.34].

An important parameter which can be determined uniquely by means of phonon echoes is the acoustic deformation potential of the two-level systems. From the strain field needed to produce maximum spontaneous echo, a value of $\gamma_L = 1.5 \pm 0.4$ eV has been obtained [7.31] for the coupling to longitudinally polarized phonons.

b) *Linewidth*

At temperatures higher than 0.1 K, information concerning T_2' comes from linewidth measurements [7.37,39,41], which are displayed in Fig.7.10. The linewidths reported [7.37] at 0.55 and 0.79 K are $\Delta\omega/2\pi$ = 45 and 65 MHz, respectively. If we assume that the spectrum of the probing pulse is broadened as much as that of the saturating pulse, the spectrum of either pulse is only half the measured width. The revised values are plotted in Fig.7.10 at $\Delta\omega/2\pi$ = 22 and 32 MHz. (The same procedure has been used for the lower temperature data [7.39].)

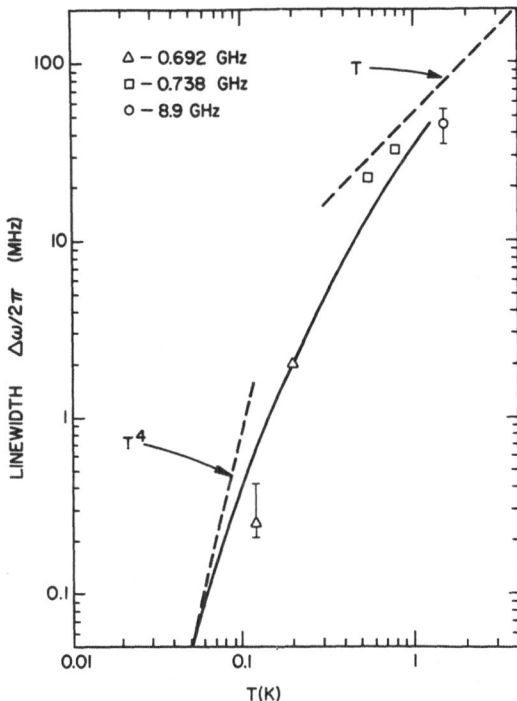

Fig.7.10. Comparison of the linewidth measured at 0.692 GHz [7.39], 0.738 GHz [7.40], and 8.9 GHz [7.41], with the linewidth (solid line) calculated from a spectral diffusion model [7.12]

A striking feature of the higher temperature data in Fig.7.10 is that the linewidth is nearly independent of frequency. At lower temperature a strong temperature dependence is observed at constant frequency.

Most of these features can be understood as the consequences of spectral diffusion. In Fig.7.10, the theoretical linewidth, which is independent of frequency, is obtained from [7.12] for a time interval of 1 μs, using γ_L = 1.4 eV to give the best agreement with all linewidths. Note that the experiments above 0.5 K were performed with simultaneous pulses. The gradual transition from the T^4 dependence (7.52) at low temperatures (T < 50 mK) to the long-time behavior proportional to T at high temperatures (T > 1 K) predicted by theory is observed. Full agreement between theory and experiment in Fig.7.10 is not expected since the two experiments

at higher temperature were performed with simultaneous pulses, whereas the calculation assumes separated pulses.

In the related, *time-dependent* linewidth experiment [7.43] described above
(Sect.7.3.3), the linewidth lies close to the calculated time-dependent linewidth.
Again, the calculation is not expected to apply quantitatively to an experiment in
which the pulse length of simultaneous pulses is varied rather than the time between
separated pulses, but the qualitative agreement appears to emphasize the importance
of spectral diffusion.

Difficulties in understanding both linewidth and saturation recovery measurements
in the presence of spectral diffusion have been pointed out previously [7.12]. Before a *quantitative* understanding of these measurements can be reached, the following problems must be addressed: a) The relationship between changes in population
and changes in attenuation is not always clear, particularly in the crossover region
between the cw and pulse regimes. b) No theory of saturation and power broadening
exists for the case in which the linewidth is time dependent on a time scale comparable to the pulse width, e.g., the transition region in Fig.7.10 between the
T and T^4 regimes.

One measurement which is not explained within the picture of spectral diffusion
is the electric echo decay [7.46,47] between 4 and 20 mK which is generally more
rapid than one would expect on the basis of spectral diffusion. In this respect we
recall the weaker-than-T^{-2} dependence of the phonon echo decay time at frequencies
high enough that $hf > 2kT$ [7.31]. We note the similarity between the low-temperature
behavior of the electric echoes and the low-temperature high-frequency trends in the
phonon echo data.

In summary, the measurements of quantities related to T_2' give support to the
prominant role of spectral diffusion. The effect of spectral diffusion on the decay
time of spontaneous phonon echoes agrees in magnitude as well as temperature dependence with the observed values, and the results of holeburning experiments at
higher temperatures generally display the temperature dependence and frequency in-
dependence expected from spectral diffusion.

7.5 Conclusions

In this review we have examined the decay rates of tunneling systems in a variety
of disordered substances and attempted to identify the processes responsible for
relaxation. Since tunneling systems may interact with phonons, photons, electrons,
and each other, the identification process involves a combination of experiment,
theory, and intuition. The exchange of energy among the various subsystems of a
glass proceeds at rates governed by matrix elements which are, a priori, not known.
Nevertheless, experimental methods exist for measuring these coupling parameters
and a certain amount of confidence exists in characterizing interaction strengths.

In all glasses examined here, the relaxation processes are dominated by their strong phonon couplings and both T_1 and T_2' reflect this extremely strong interaction. The only possible exception occurs in metallic glasses where a larger, but parallel, relaxation rate exists as a result of a tunneling system-conduction electron interaction.

To conclude, we must acknowledge that important questions remain unanswered in understanding the dynamics of tunneling systems and, therefore, we suggest the following areas for future research.

1) T_1 processes-- Although we have asserted that (7.38) is valid, no direct test of the E-dependence has been made. Above a few K, one should expect higher-order phonon processes to appear, but no experimental evidence has demonstrated this. Similarly, in a metallic glass no direct test of (7.46) has yet been performed.

2) T_2 processes-- Spectral diffusion through phonon interactions has provided a reasonable explanation of T_2', or $\Delta\omega$ in insulators, but some inconsistencies exist. Foremost is the apparent absence of Gaussian two-pulse echo decay at short times. The frequency dependence of the phonon echo decay is, as yet, unexplained. Below 20 mK, T_2' does not appear to vary as T^{-2} and, more significantly, no τ_{12} dependence of the stimulated electric echo is observed. It may be that another, more effective, dephasing process occurs as the temperature is lowered or as the frequency is increased. Finally, the large electric dipoles of OH-associated tunneling centers should produce dephasing rates due to electric dipolar coupling comparable to phonon coupling. This effect has not been observed.

References

7.1 R.C. Zeller, R.O. Pohl: Phys. Rev. B4, 2029 (1971)
7.2 P.W. Anderson, B.I. Halperin, C.M. Varma: Philos. Mag. 25, 1 (1972)
7.3 W.A. Phillips: J. Low Temp. Phys. 7, 351 (1972)
7.4 S. Hunklinger, W. Arnold, S. Stein, R. Nava, K. Dransfeld: Phys. Lett. 42A, 253 (1972)
7.5 B. Golding, J.E. Graebner, B.I. Halperin, R.J. Schutz: Phys. Rev. Lett. 30, 223 (1973)
7.6 A. Abragam: *The Principles of Nuclear Magnetism* (Oxford University Press, London 1961)
7.7 R.P. Feynman, F.L. Vernon, R.W. Hellwarth: J. Appl. Phys. 28, 49 (1957)
7.8 C.P. Slichter: *Principles of Magnetic Resonance*, Springer Series in Solid-State Sciences, Vol.1, 2nd ed. (Springer, Berlin, Heidelberg, New York 1980)
7.9 R.H. Pantell, H.E. Puthoff: *Fundamentals of Quantum Electronics* (Wiley, New York 1969)
7.10 M. Sargent, M.O. Scully, W.E. Lamb, Jr.: *Laser Physics* (Addison-Wesley, Reading, MA 1974)
7.11 J. Jäckle: Z. Phys. 257, 212 (1972)
7.12 J.L. Black, B.I. Halperin: Phys. Rev. B16, 2879 (1977)
7.13 B. Golding, J.E. Graebner, W.H. Haemmerle: In *Proceedings Int. Conf. on Lattice Dynamics*, Paris (1977), ed. by M. Balkanski (Flammarion, Paris 1978) p.348
7.14 B. Golding, J.E. Graebner, A.B. Kane, J.L. Black: Phys. Rev. Lett. 41, 1487 (1978)

7.15 B.I. Halperin: Ann. N.Y. Acad. Sci. *279*, 173 (1976)
7.16 J. Joffrin, A. Levelut: J. Phys. Paris *36*, 811 (1975)
7.17 J.L. Black: Phys. Rev. B*17*, 2740 (1978)
7.18 S. Hunklinger, L. Pichê: Solid State Commun. *17*, 1189 (1975)
7.19 B. Golding, J.E. Graebner, R.J. Schutz: Phys. Rev. B*14*, 1660 (1976)
7.20 R. Maynard: In *Phonon Scattering in Solids*, ed. by L.J. Challis, V.W. Rampton, A.F.G. Wyatt (Plenum, NY 1976) p.115
7.21 G. Bellessa, P. Doussineau, A. Levelut: J. Phys. Paris Lett. *38*, L-65 (1977)
7.22 J.R. Matey, A.C. Anderson: Phys. Rev. B*16*, 3406 (1977)
7.23 J.E. Graebner, B. Golding, R.J. Schutz, F.S.L. Hsu, H.S. Chen: Phys. Rev. Lett. *39*, 1480 (1977)
7.24 J. Kondo: Physica Utrecht *40*B 40, 207 (1976)
7.25 J.L. Black, B.L. Gyorffy: Phys. Rev. Lett. *41*, 1595 (1978)
7.26 J. Korringa: Physica Utrecht *16*, 601 (1950)
7.27 J.R. Klauder, P.W. Anderson: Phys. Rev. *125*, 912 (1962)
7.28 W.B. Mims: In *Electron Paramagnetic Resonance*, ed. by S. Geschwind (Plenum, New York 1972)
7.29 B. Golding, J.E. Graebner: Phys. Rev. Lett. *37*, 852 (1976)
7.30 B. Golding, J.E. Graebner, W.H. Haemmerle: In *Amorphous and Liquid Semiconductors*, ed. by W.E. Spear (CICL University of Edinburgh 1977) p.367
7.31 J.E. Graebner, B. Golding: Phys. Rev. B*19*, 964 (1979)
7.32 E.L. Hahn: Phys. Rev. *80*, 580 (1950)
7.33 I.D. Abella, N.A. Kurnit, S.R. Hartmann: Phys. Rev. *141*, 391 (1966)
7.34 P. Hu, L.R. Walker: Solid State Commun. *24*, 813 (1977)
7.35 P.F. Liao, S.R. Hartmann: Phys. Rev. B*8*, 69 (1973)
7.36 B. Golding, M. v. Schickfus, S. Hunklinger, K. Dransfeld: Phys. Rev. Lett. *43*, 1817 (1979)
7.37 S. Hunklinger, W. Arnold: In *Physical Acoustics*, Vol.XII, ed. by W.P. Mason, R.N. Thurston (Academic, New York 1976) p.155
7.38 J.E. Graebner, B. Golding: Bull. Am. Phys. Soc. *22*, 310 (1977)
7.39 B. Golding, J.E. Graebner: Unpublished
7.40 W. Arnold, S. Hunklinger: Solid State Commun. *17*, 883 (1975)
7.41 A. Bachellerie, P. Doussineau, A. Levelut, T.-T. Ta: J. Phys. Paris *38*, 69 (1977)
7.42 B. Golding: IEEE Trans. SU-*24*, 692 (1977)
7.43 W. Arnold, C. Martinon, S. Hunklinger: J. Phys. Paris Lett. *39*, C6-961 (1978)
7.44 L. Bernard, L. Pichê, G. Schumacher, J. Joffrin, J.E. Graebner: J. Phys. Paris Lett. *39*, L-126 (1978)
7.45 M. v. Schickfus, B. Golding, W. Arnold, S. Hunklinger: J. Phys. Paris *39*, C6-959 (1978)
7.46 L. Bernard, L. Pichê, G. Schumacher, J. Joffrin: J. Low Temp. Phys. *35*, 411 (1979)
7.47 L. Pichê: J. Phys. Paris *39*, C6-1545 (1978)
7.48 P.J. Anthony, A.C. Anderson: Phys. Rev. B*20*, 763 (1979)
7.49 G. Frossati, J. le G. Gilchrist, J.C. Lasjaunias, W. Meyer: J. Phys. C*10*, L515 (1977)

Additional References

B. Golding, J.E. Graebner, W.H. Haemmerle: "Microwave Photon Echos from Polyethylene", Phys. Rev. Lett. *44*, 899 (1980)

8. Low Frequency Raman Scattering in Glasses

J. Jäckle

With 10 Figures

The subject of this chapter is the inelastic scattering of light in glasses with frequency shifts smaller than about 1/5 of the Debye frequency. This field is commonly referred to as low-frequency Raman scattering in glasses. The term Raman scattering is used here for the whole spectrum of the scattered light with the exception only of the distinct Brillouin lines.

8.1 Introductory Comments

The low-frequency Raman spectrum consists of two parts: The low-frequency tail of the first-order vibrational spectrum and a broad quasielastic line which extends beyond the Brillouin lines. As for the quasielastic line, the term Rayleigh scattering would be more adequate, but because of the overlap with the first-order vibrational contribution this distinction is not normally made. When dealing with glasses, the term Rayleigh scattering is used only for the strictly elastic contribution from the static optical inhomogeneity. In this research two major points of interest are 1) to determine the asymptotic low-frequency form of the first-order vibrational Raman tensor and to interpret the deviations from this form, which occur at intermediate frequencies, in terms of dispersion and short-range atomic order, and 2) to explain the origin of the quasielastic scattering and to understand how it is related to the disordered atomic structure.

8.2 Vibrational Raman Spectrum of First Order

8.2.1 Experimental Results

Raman scattering in amorphous semiconductors has been reviewed by BRODSKY [8.1] and LUCOVSKY [8.2]. The key to an understanding of the experimental Raman spectra is the breakdown of the wave-vector selection rule of crystalline Raman scattering. This leads to a continuous first-order vibrational spectrum of glasses instead of the discrete Raman spectrum of crystals (Fig.8.1). Two properties of glass are responsible for this fundamental difference. Firstly, in glass the coupling

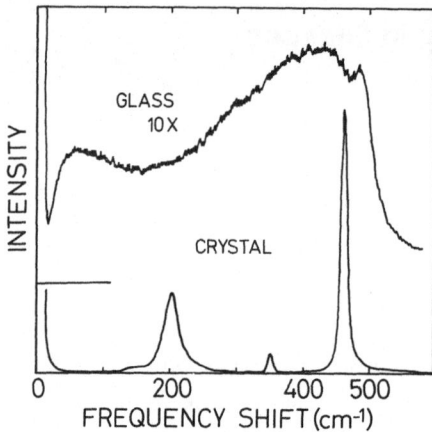

INTENSITY

GLASS
10X

CRYSTAL

0   200   400 500
FREQUENCY SHIFT (cm⁻¹)

<u>Fig.8.1.</u> Comparison of the Raman spectra of vitreous and crystalline silica [8.3b]

between the atomic displacement and the fluctuation of the dielectric susceptibility is itself a random quantity because of the irregular atomic bonding. This aspect of the disorder is referred to as "electrical disorder". Secondly, with the exception of very low frequencies, the vibrational modes of the glass are not plane waves. The disorder which manifests itself by the distortion of the vibrational modes is termed "mechanical disorder". Both kinds of disorder also exist in impurity-doped crystals, where the wave-vector selection rule is partially broken.

The effect of the electrical disorder on the Raman scattering in molecular crystals with random molecular orientation has been studied by WHALLEY and BERTIE [8.4]. Even for a perfect plane-wave phonon mode with finite wave vector $\vec{q}$, the disorder of the atomic coupling allows inelastic scattering of light from this mode. With the $(-\vec{q})$-component of the random spatial distribution of atomic couplings, the phonon $\vec{q}$ is folded back to yield a $\vec{k} \approx 0$ component of the fluctuating dielectric polarizability which can be measured by light scattering.

The effect of the mechanical disorder on the Raman scattering was first explained by SHUKER and GAMON [8.3]. They define a correlation length $\xi(\omega)$ over which a vibrational mode of frequency $\omega/2\pi$ represents a plane wave and argue that the wave-vector selection rule is annulled as soon as $\xi(\omega)$ is as short as the mode wavelength.

The intensity $I(\omega)$ of light scattering with a frequency shift $(-\omega/2\pi)$ is determined by the Raman tensor $I_{ij}(\omega)$ defined as the Fourier transform of the correlation function for the space- and time-dependent fluctuations of the dielectric susceptibility χ.

$$I_{ij}(\omega) = 1/(2\pi\Omega) \int_{-\infty}^{+\infty} dt \, e^{i\omega t} \int d^3r_1 d^3r_2 \, e^{-i\vec{k}(\vec{r}_1-\vec{r}_2)}$$

$$\cdot \langle \chi_i(\vec{r}_1,t)\chi_j(\vec{r}_2,0)\rangle \quad (i,j = 1, \ldots 6) \quad , \tag{8.1}$$

where Ω is the volume of the system, the brackets <...> denote the equilibrium expectation value, and the Voigt notation for tensor components is used. Since the wave-vector transfer $\vec{k}$ from the scattered light is small, for an amorphous system the Raman tensor is independent of $\vec{k}$ except for the contribution of the Brillouin spectrum. For the Raman spectrum k may be put equal to zero in (8.1). As a result of the electrical and mechanical disorder, the entire vibrational density of states $g(\omega)$ contributes to the Raman scattering in glasses. The expression for the disorder-induced vibrational contribution to the Raman tensor can be written, as shown by SHUKER and GAMON [8.3],

$$I_{ij}(\omega) = C_{ij}(\omega)g(\omega)[1 + n(\omega)]/\omega \quad . \tag{8.2}$$

Here $n(\omega)$ denotes the Bose distribution function $n(\omega) = [\exp(\hbar\omega/k_B T) - 1]^{-1}$ for temperature T. $C_{ij}(\omega)$ describes the coupling of the vibrational modes of frequency $\omega/2\pi$ to the light for a particular polarization geometry. In the extreme case of complete randomness of the atomic couplings or the atomic amplitudes in the vibrational modes, $C_{ij}(\omega)$ would be independent of frequency. In this case the reduced Raman tensor, defined as

$$I^R_{ij}(\omega) = I_{ij}(\omega)\omega/[1 + n(\omega)] \quad , \tag{8.3}$$

would be proportional to the density of states $g(\omega)$ with a constant proportionality factor.

The temperature dependence of the Raman intensity (8.2) proportional to $[1 + n(\omega)]$ is the signature of a first-order process. It has been amply verified experimentally. This factor explains the different form of the measured room-temperature and low-temperature Raman spectra. At room temperature the Raman intensity exhibits a peak at lower frequencies in the region of the acoustic phonons of the corresponding crystal (at 52 cm^{-1} in vitreous silica and at about 150 cm^{-1} in a-Si). In several cases this peak disappears at low temperatures ($k_B T \ll \hbar\omega$), for example, in the polarized Raman spectrum of vitreous silica (Fig.8.2). The existence of a peak in the room-temperature Raman spectrum, in a frequency region where $\hbar\omega \ll k_B T$, implies a) that the factor

$$C_{ij}(\omega)g(\omega)/\omega^2 = I^R_{ij}(\omega)/\omega^2 \tag{8.4a}$$

in (8.2) for the Raman tensor exhibits the same maximum. If the peak disappears at low temperatures, on the other hand, it can be concluded b) that the function

$$\omega C_{ij}(\omega)g(\omega)/\omega^2 = I^R_{ij}(\omega)/\omega \tag{8.4b}$$

increases monotonously with increasing frequency. Because of its connection with the Bose distribution function, the room-temperature peak in the Raman spectrum at low frequencies is often referred to as the "boson peak". If the Raman coupling $C_{ij}(\omega)$ were independent of frequency, the boson peak at room temperature would be

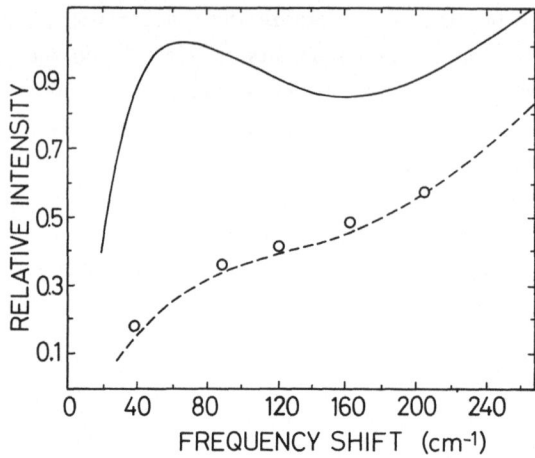

Fig.8.2. Comparison of measured polarized spectrum of vitreous si-lica at 14 K with that computed using the room-temperature inten-sity [8.5]. ———— 300 K data, ----- 300 K data reduced to 14 K, O 14 K data

entirely due to a maximum in the reduced density of states $g(\omega)/\omega^2$. This assump-tion leads to an interpretation of the low-frequency Raman spectrum which is simi-lar to that given earlier by FLUBACHER et al. [8.6]. They inferred the existence of extra vibrational modes in vitreous silica by which the excess heat capacity at low temperatures (3 - 15 K) over the Debye contribution could be explained. However, the Raman coupling is strongly frequency dependent in this frequency range and, according to MARTIN and BRENIG [8.7], also contributes to the appearance of the boson peak.

Beyond the range of the vibrational density of states, any Raman scattering must be due to higher-order processes. Such is the case, for example, in a-Si for $600 \leq \bar{\omega} \leq 1050 \text{ cm}^{-1}$ [8.8] and in vitreous silica for $1400 \leq \bar{\omega} \leq 2600 \text{ cm}^{-1}$ [8.9]. Similarly, two-phonon processes dominate over the one-phonon process in spectral regions where the vibrational density of states $g(\omega)$ is very low.

The reduced Raman tensor $I_{ij}^{R}(\omega)$ (8.3), for various polarizations, can be com-pared with the vibrational density of states $g(\omega)$ obtained from inelastic neutron scattering or model calculations (see Chap.2). In general, a considerable similarity between the two curves is found, but the low-frequency part of the density of states is generally less well represented in the reduced Raman intensity (Fig.8.3). The coupling $C_{ij}(\omega)$, therefore, is not constant, but generally decreases with decreas-ing frequency. In addition, the coupling may depend fairly strongly on the parti-cular polarization geometry. For vitreous silica (Fig.8.4) the relative reduction of the low-frequency part of the Raman spectrum is much stronger in the polarized component (VV) than in the depolarized component (VH). For a more detailed account the reader is referred to Brodsky's review article [8.1].

At the lower end of the Raman spectrum (between about 30 - 35 and 67 cm^{-1}) the Raman coupling $C_{ij}(\omega)$ of a-Si was found to be proportional to ω^2 (Fig.8.5) [8.11]. A similar behaviour was observed for a-As$_2$S$_3$, a-GeS$_2$ and a-GeSe$_2$ [8.12]. There are good theoretical arguments, to be described below, that this should be the limiting

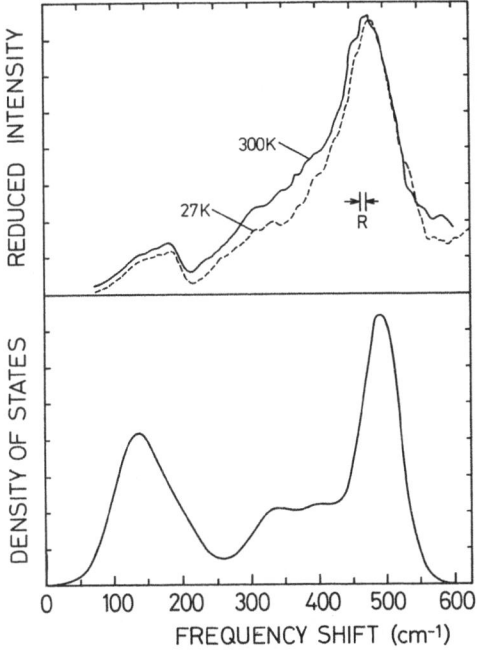

Fig.8.3. Comparison of Raman spectrum and vibrational density of states g(ω) of a-Si. The density of states curve is obtained by convoluting the crystalline g(ω) with a Gaussian of appropriate width [8.8]

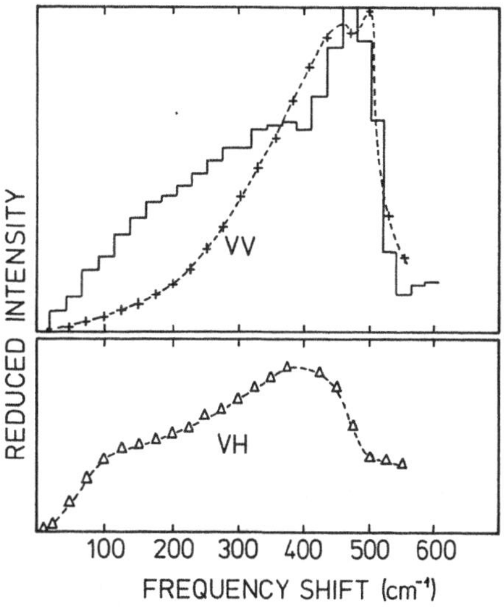

Fig.8.4. Reduced polarized (VV) and depolarized (VH) Raman intensities of vitreous silica [8.3]. The histogram is the density of states calculated by BELL et al. [8.10]

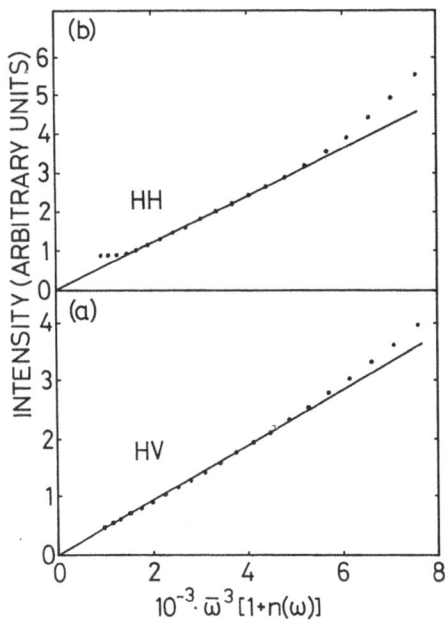

Fig.8.5. Raman intensities of a-Si at low frequencies [8.11] (ω: frequency shift in cm^{-1})

behavior of the first-order Raman coupling of the vibration modes of low frequency. In Raman data for vitreous silica [8.13] the limiting low-frequency behavior of the vibrational Raman coupling was somewhat obscured by the overlap with
the broad quasielastic component of the light scattering spectrum.

8.2.2 Theory of Low-Frequency Spectrum

a) *Continuum Theory*

For the treatment of the first-order vibrational Raman scattering at very low frequencies a continuum description is appropriate [8.7]. Before turning to the more
specific results of Martin and Brenig's theory, we first show that the contributions
to the Raman coupling $C_{ij}(\omega)$ from electrical and mechanical disorder both vary as
ω^2 in the limit of low frequencies. Using data on the mean free path of thermal
phonons at low temperatures one can estimate that the contribution from electrical
disorder dominates for vitreous silica.

The contribution of a vibrational mode λ to the Raman tensor can be expressed
in terms of the coupling

$$\tilde{C}_{ij}(\lambda) = (1/\Omega) \int d^3r_1 d^3r_2 \delta x_i(\lambda,\vec{r}_1)\delta x_j^*(\lambda,\vec{r}_2) \qquad (i,j = 1, \ldots 6) \quad , \qquad (8.5)$$

where $\delta x_i(\lambda,\vec{r}_1)$ is the local fluctuation of the component x_i of the dielectric
susceptibility tensor at position $\vec{r}_1$ which is caused by the vibration mode λ. In
a continuum description the fluctuations $\delta x_i(\lambda,\vec{r})$ result from the elastic strain
field $e_j(\lambda,\vec{r})$ associated with mode λ, which is modulated by static fluctuations
$\delta p_{ij}(\vec{r})$ of the elasto-optical constants,

$$\delta x_i(\lambda,\vec{r}) = -(\epsilon^2/4\pi)[p_{ik} + \delta p_{ik}(\vec{r})]e_k(\lambda,\vec{r}) \quad . \qquad (8.6)$$

The fluctuations $\delta p_{ij}(\vec{r})$ due to the elastooptical inhomogeneity of the glass represent the "electrical disorder". Neglecting at first the mechanical disorder,
one can describe the vibration λ as a plane wave of wave vector $\vec{q}_\lambda$ with a polarization vector $\vec{u}(\lambda)$. The strain field $e_j(\lambda,\vec{r})$ can be written as

$$e_k(\lambda,\vec{r}) = q_\lambda e_k(\lambda)e^{i\vec{q}_\lambda \cdot \vec{r}} \qquad (8.7)$$

where

$$e_k(\lambda) = e_{\alpha\beta}(\lambda) = (1/2)[u_\alpha(\lambda)\hat{q}_\beta + \hat{q}_\alpha u_\beta(\lambda)] \quad , \qquad (\hat{q} = \vec{q}_\lambda/q_\lambda) \quad ,$$

and k is the Voigt notation for the Cartesian ($\alpha\beta$) tensor component. In terms of
the correlation functions of the fluctuations $\delta p_{ij}(\vec{r})$ of the elastooptical constants

$$\overline{\delta p_{ij}(\vec{r})\delta p_{kl}(0)} = (1/\Omega) \int d^3r_1 d^3r_2 \delta(\vec{r}_1 - \vec{r}_2 - \vec{r})\delta p_{ij}(\vec{r}_1)\delta p_{kl}(\vec{r}_2) \quad , \qquad (8.8)$$

which characterize the electrical disorder, the coupling $\tilde{c}_{ij}^{(e)}(\lambda)$ is given by

$$\tilde{c}_{ij}^{(e)}(\lambda) = (\varepsilon^2/4\pi)^2 q_\lambda^2 e_k(\lambda) e_1^*(\lambda) \int d^3r \, e^{i\vec{q}_\lambda \cdot \vec{r}} \, \overline{\delta p_{ik}(\vec{r})\delta p_{j1}(0)} \quad . \tag{8.9}$$

For acoustic phonon modes with $\omega(\lambda) = c_\lambda \cdot q_\lambda$ one obtains

$$\tilde{c}_{ij}^{(e)}(\lambda) \propto \omega(\lambda)^2 \quad \text{for} \quad \omega(\lambda) \to 0 \quad . \tag{8.10}$$

This result has been derived by PRETTL et al. [8.14] from a rather elementary argument.

The mechanical disorder, defined as the irregularity of the form of the vibrational modes, is characterized by the correlation function for the irregular local strain field $e_j(\lambda,\vec{r})$ of the mode,

$$\overline{e_k(\lambda,\vec{r})e_1^*(\lambda,0)} = (1/\Omega) \int d^3r_1 d^3r_2 \delta(\vec{r}_1 - \vec{r}_2 - \vec{r}) e_k(\lambda,\vec{r}_1) e_1^*(\lambda,\vec{r}_2) \quad . \tag{8.11}$$

It is assumed that the scattering of a plane wave of wave vector $\vec{q}_\lambda$ in the inhomogeneous medium gives rise to an exponential damping factor $\exp(-r/\xi_\lambda)$ in this correlation function ([8.3], for a detailed analysis see also [8.15]), which then reads

$$\overline{e_k(\lambda,\vec{r})e_1^*(\lambda,0)} = q_\lambda^2 e_k(\lambda) e_1^*(\lambda) \exp(i\vec{q}_\lambda \cdot \vec{r} - r/\xi_\lambda) \quad . \tag{8.12}$$

This expression yields the following contribution from mechanical disorder to the coupling $\tilde{C}_{ij}(\lambda)$:

$$\tilde{c}_{ij}^{(m)}(\lambda) = (\varepsilon^2/4\pi)^2 q_\lambda^2 e_k(\lambda) e_1^*(\lambda) \frac{8\pi\xi_\lambda^{-1}}{(q_\lambda^2 + \xi_\lambda^{-2})^2} p_{ik} p_{j1} \quad . \tag{8.13}$$

This is essentially the result of SHUKER and GAMON. It is now further assumed that the correlation length ξ_λ is related to the phonon mean free path, which arises from the scattering of a plane wave in the irregular structure, by

$$\xi_\lambda^{-1} = (1/2) l_\lambda^{-1} \quad . \tag{8.14}$$

At lower frequencies where the condition $q_\lambda \xi_\lambda \gg 1$ is fulfilled and a wave vector $\vec{q}_\lambda$ can be defined, this phonon mean free path follows the Rayleigh scattering law for long-wavelength acoustic phonons

$$l_\lambda^{-1} = a q_\lambda^4 \quad , \tag{8.15}$$

where the constant a may depend on the polarization branch. With the expression (8.13) for $\tilde{c}_{ij}^{(m)}(\lambda)$, this result leads to the same quadratic frequency dependence of the coupling due to the mechanical disorder at low frequency as in the case of the electrical disorder!

In the limit of low frequency the ratio of the couplings arising from either of the two kinds of disorder is given by

$$\frac{\tilde{C}_{ij}^{(m)}(\lambda)}{\tilde{C}_{ij}^{(e)}(\lambda)} = \frac{4\pi a p_{ik} p_{j1} \cdot e_k(\lambda) e_1^*(\lambda)}{\int d^3 r \, \overline{\delta p_{ik}(\vec{r}) \delta p_{j1}(0)} \cdot e_k(\lambda) e_1^*(\lambda)} \quad . \tag{8.16}$$

It can be shown that this ratio is of the order of unity if only the contribution of the static density fluctuations to the phonon Rayleigh scattering and to the correlation function of the fluctuating elastooptical constants are taken into account. These contributions are related to the correlation function $\overline{\delta p(\vec{r}) \delta p(0)}$ of the static density fluctuations in the following way:

$$1_\lambda^{-1}\Big|_{\delta p} = a\Big|_{\delta p} q_\lambda^4 = (q_\lambda^4/4\pi)\gamma_c^2 \int d^3 r \, \overline{\delta p(\vec{r}) \delta p(0)}/\rho_0^2$$

$$\int d^3 r \, \overline{\delta p_{ik}(r) \delta p_{j1}(0)}/p_{ik} p_{j1}\Big|_{\delta p} = \gamma_p^2 \int d^3 r \, \overline{\delta p(\vec{r}) \delta p(0)}/\rho_0^2 \quad . \tag{8.17}$$

γ_c is an average of Grüneisen parameters for the sound velocities. For vitreous silica γ_c^2 has a value of 24 and 7 for longitudinal and transverse waves, respectively [8.16]. $\gamma_p = \partial \ln p/\partial \ln \rho$ is an average Grüneisen parameter for the elasto-optical constants. For vitreous silica γ_p^2 can be estimated to be of the same order of magnitude as the value of γ_c^2 [8.17]. Consequently, for the contributions from the static density fluctuations

$$\frac{\tilde{C}_{ij}^{(m)}(\lambda)}{\tilde{C}_{ij}^{(e)}(\lambda)}\Big|_{\delta p} \approx \frac{\gamma_c^2}{\gamma_p^2} = O(1) \quad \text{for} \quad \omega(\lambda) \to 0 \quad . \tag{8.18}$$

The total coefficient a in (8.15), obtained from experimental phonon mean free path data, is about two orders of magnitude larger than the contribution calculated from the known static density fluctuations in vitreous silica [8.16]. For the same material the total coupling $C_{ij}(\omega)$, however, is about 3000 times larger than the contribution of the static density fluctuations to the electrical disorder [8.18]. From these two results it can be concluded that the total contribution of the electrical disorder to the coupling $C_{ij}(\omega)$ exceeds the contribution of the mechanical disorder by at least one order of magnitude. This estimate holds on average for different polarization geometries. For individual tensor components of the coupling the ratio of the mechanical and the electrical contribution may, of course, deviate from the average.

MARTIN and BRENIG [8.7] did not use the Shuker and Gamon ansatz (8.12) for the correlation function of the local strain field in their treatment of the mechanical disorder. They described the vibrations as plane-wave phonons in a reference frame relative to which the elastic medium is locally displaced by a frequency-independent vector $\vec{A}(\vec{r})$. Correlation functions for such distorted plane-wave phonons are calculated by expanding to second order in the displacement field $\vec{A}(\vec{r})$. The cross terms of mechanical and electrical disorder are estimated to be negligible. For low frequencies a quadratic frequency dependence of the Raman coupling $C_{ij}(\omega)$ is also obtained. The necessary correction of this result at intermediate frequencies is derived from an assumed spatial decay of the correlation function for the static fluctuation (8.8) of the elastooptical constants. For the Martin and Brenig assumption of a Gaussian decay, viz.,

$$\overline{\delta p_{ij}(\vec{r})\delta p_{k1}(0)} = \overline{\delta p_{ij}\delta p_{k1}} \ \exp[-r^2/(2\sigma)^2] \ , \tag{8.19}$$

the Raman coupling of acoustic phonons of polarization branch i is proportional to

$$C_{k1}(\omega)\Big|_i \propto (\omega/c_i)^2 \ \exp[-(\omega\sigma/c_i)^2] \ , \tag{8.20}$$

according to (8.9). Here c_i is the sound velocity for polarization i. MARTIN and BRENIG argued that the correlation length 2σ is determined by the short-ranged atomic order as reflected in the radial distribution function and is typically of the order of 5 - 10 Å. Equation (8.20) has proved useful for a fit of experimental Raman data since it reproduces the observed low-frequency maximum in the room-temperature Raman spectrum. Combined with a Debye density of states $g(\omega)$, this formula leads to a position of the maximum close to $\omega = c_t/\sigma$, where c_t is the transverse sound velocity. Following the Martin and Brenig interpretation, the values of the correlation length 2σ obtained from such fits have been used a measures of the extension of the short-range atomic order in various glasses [8.12,19,20]. This application of the Martin and Brenig theory should, however, be considered with some caution for two reasons. First, the Debye approximation of the vibrational density of states is inaccurate in the frequency region where the reduced Raman intensity deviates from the limiting ω^4 behavior [8.11]. The deficiency of the Debye model is related to the fact that the acoustic phonon branches of the corresponding crystal show strong dispersion in the same frequency region. In general, it may be expected that the variation of the factor $g(\omega)/\omega^2$, which is caused by the deviation from a Debye density of states, is of similar importance for the form of the Raman spectrum as the deviation of the Raman coupling $C_{ij}(\omega)$ from a quadratic frequency dependence. Secondly, the Gaussian ansatz (8.19) for the correlation function appears somewhat arbitrary. Obviously, the number obtained for the correlation length from a fit of Raman data depends on the particular choice of the analytical expression for this correlation function. Moreover, it may be noted that the assumed strong decay of the coupling function (8.20) automatically leads to a maximum in the low-temperature Raman spectrum, which is, however, not observed in several cases.

As a further result of their theory, MARTIN and BRENIG [8.7] obtained the correct depolarization ratio for the back scattering experiment with a-Si [8.11]. Assuming that the fluctuations $\overline{(\delta p_{ij})^2}$ of the elastooptical constants are proportional to the square p_{ij}^2 of their average values, viz.,

$$\overline{(\delta p_{ij})^2} = \lambda p_{ij}^2 \tag{8.21}$$

with λ independent of i and j, they obtained

$$I_{HH}/I_{HV} \approx 1.5 \quad, \tag{8.22}$$

which agrees with the experimental value

$$I_{HH}/I_{HV}\Big|_{exp} = 1.5 \pm 0.2 \quad. \tag{8.23}$$

Here H and V refer to light polarized parallel and perpendicular to the scattering plane, which in this case is defined by the wave vector of the light and the normal to the surface. WINTERLING [8.13], applying the same theory to the low-frequency Raman scattering in vitreous silica, calculated the depolarization ratio in 90° scattering to be

$$I_{HV}/I_{VV} \approx 0.28 \quad. \tag{8.24}$$

His experimental value for the frequency region 6 - 40 cm^{-1} is

$$I_{HV}/I_{VV}\Big|_{exp} = 0.30 \pm 0.03 \tag{8.25}$$

and confirms within experimental error the theoretical result. However, according to NEMANICH [8.12], for the chalcogenide glasses As_2S_3, GeS_2, and $GeSe_2$ the calculated depolarization ratios I_{HV}/I_{VV} are too small by more than a factor 2.

b) *Bond Polarizability Model*

So far the theory of the low-frequency vibrational Raman spectrum of first order has been based completely on a continuum description in which the inhomogeneity of elastic and elastooptical properties characterizes a glass. In an atomistic theory the electrical disorder is described by the irregular displacement-dependent atomic polarizabilities, which correspond to the elastooptical constants in the continuum description. A relatively simple model for these polarizabilities, which applies to materials with covalent bonding, is the bond polarizability model [8.21,22]. The polarizability $\alpha_{\beta\beta'}^{(nm)}$ of a bond between nearest neighbors n and m is given by

$$\alpha_{\beta\beta'}^{(nm)} = \alpha(r_{nm})\delta_{\beta\beta'} + \gamma(r_{nm})[\hat{r}_{nm,\beta}\hat{r}_{nm,\beta'} - (1/3)\delta_{\beta\beta'}] \quad, \tag{8.26}$$

where $\vec{r}_{nm}$ is the nearest neighbor relative position vector and $\hat{r}_{nm}$ the corresponding unit vector. Expanding this expression with respect to the relative displacement

$$\vec{u}_{nm} = \vec{r}_{nm} - \vec{R}_{nm} \quad ,$$

where $\vec{R}_{nm}$ is the relative position vector in equilibrium, one obtains the follow-ing expression for the displacement-dependent polarizability of first order:

$$\alpha_{\beta\beta'}^{(nm)}(\vec{u}_{nm}) = \alpha'(\vec{u}_{nm} \cdot \hat{R}_{nm})\delta_{\beta\beta'} + \gamma'(\vec{u}_{nm} \cdot \hat{R}_{nm})[\hat{R}_{nm,\beta}\hat{R}_{nm,\beta'} - (1/3)\delta_{\beta\beta'}]$$

$$+ \gamma R_{nm}^{-1}[u_{nm,\beta}\hat{R}_{nm,\beta'} + \hat{R}_{nm,\beta}u_{nm,\beta'} \tag{8.27}$$

$$- 2(\vec{u}_{nm} \cdot \hat{R}_{nm})\hat{R}_{nm,\beta}\hat{R}_{nm,\beta'}]$$

where

$$(\hat{R}_{nm} = \vec{R}_{nm}/R_{nm}) \quad .$$

The total polarizability is the sum over all pairs of atoms (nm). This form of the displacement-dependent polarizability, which contains the three parameters α', γ, and γ', has been used to calculate the vibrational Raman spectra of a-Si and a-Ge [8.21], a-As and a-Se [8.23], a-SiO$_2$ and the homologues a-GeO$_2$ and a-BeF$_2$ [8.22], and of a-B$_2$O$_3$ and mixed SiO$_2$-B$_2$O$_3$ glasses [8.24]. For low frequencies where the vibrational modes represent acoustic phonons with a finite mean free path, the quadratic frequency dependence of the Raman couplings $C_{ij}(\omega)$ follows readily from the displacement-dependent polarizability (8.27), since for wavelengths $2\pi/q_\lambda$ large compared to interatomic distances the relative displacement is proportional to

$$\vec{u}_{nm} \propto (e^{i\vec{q}_\lambda \cdot \vec{R}_n} - e^{i\vec{q}_\lambda \cdot \vec{R}_m}) \propto i\vec{q}_\lambda \cdot \vec{R}_{nm} \quad . \tag{8.28}$$

Neglecting mechanical disorder and assuming plane-wave acoustic phonon modes λ with wave vector $\vec{q}_\lambda$ and polarization vector $\vec{u}(\lambda)$, for this bond polarizability model the coupling $\tilde{c}_{ij}^{(e)}(\lambda)$ can be expressed as follows (returning to the Voigt notation):

$$\tilde{c}_{ij}^{(e)}(\lambda) = (1/\Omega) \sum_{(nm),(n'm')} \alpha_i^{(nm)}[\vec{u}_{nm}(\lambda)] \cdot \alpha_j^{(n'm')}[\vec{u}_{n'm'}(\lambda)]$$

$$= n_b \int d^3r \, e^{i\vec{q}_\lambda \cdot \vec{r}}[g_b(r) + \delta(\vec{r})]S_{ij}(\lambda,\vec{r}) \quad . \tag{8.29}$$

Here $g_b(r)$ is the radial distribution function of the bonds with the bond position defined by the bond center

$$g_b(r) = (1/N_b) \sum_{(nm)\neq(n'm')} \delta\left(\frac{\vec{R}_n + \vec{R}_m}{2} - \frac{\vec{R}_{n'} + \vec{R}_{m'}}{2} - r\right) , \tag{8.30}$$

and N_b is the total number of bonds. $n_b = N_b/\Omega$ is the average bond density. The function $S_{ij}(\lambda,\vec{r})$ is the conditional correlation function of the mode-dependent polarizabilities $\alpha_{\beta\beta'}^{(nm)}(\lambda)$ given by

$$
\begin{aligned}
\alpha_{\beta\beta'}^{(nm)}(\lambda) &= \alpha_{\beta\beta'}^{(nm)}[\vec{u}_{nm}(\lambda)] \cdot \exp[-i\vec{q}_\lambda(\vec{R}_n + \vec{R}_m)/2] \\
&= 2i \sin(\vec{q}_\lambda \cdot \vec{R}_{nm}/2)\left(\alpha'[\vec{u}(\lambda) \cdot \hat{R}_{nm}]\delta_{\beta\beta'}\right. \\
&\quad + \gamma'[\vec{u}(\lambda) \cdot \hat{R}_{nm}][\hat{R}_{nm,\beta}\hat{R}_{nm,\beta'} - (1/3)\delta_{\beta\beta'}] \\
&\quad + \left.\gamma R_{nm}^{-1}\{u_\beta(\lambda)\hat{R}_{nm,\beta'} + R_{nm,\beta}u_{\beta'}(\lambda) - 2[\vec{u}(\lambda) \cdot \hat{R}_{nm}]\hat{R}_{nm,\beta}\hat{R}_{nm,\beta'}\}\right) .
\end{aligned}
\tag{8.31}
$$

The condition under which the correlation function has to be evaluated is that a pair (nm), (n'm') of bonds in the irregular network is separated by $\vec{r}$. In other words, $S_{ij}(\lambda,\vec{r})$ is defined as the average of the product

$$
\alpha_i^{(nm)}(\lambda)\left[\alpha_j^{(n'm')}(\lambda)\right]^* \tag{8.32}
$$

taken over all pairs of bonds with

$$
\frac{\vec{R}_n + \vec{R}_m}{2} - \frac{\vec{R}_{n'} + \vec{R}_{m'}}{2} = \vec{r} .
$$

The expression (8.29) can be rewritten in terms of the Fourier transform

$$
S_{ij}(\lambda,\vec{k}) = \int d^3r \, e^{i\vec{k}\vec{r}} S_{ij}(\lambda,\vec{r}) \tag{8.33}
$$

and the "bond structure factor"

$$
S_b(k) = (1/N_b) \sum_{(nm)\neq(n'm')} \exp[i\vec{k} \cdot (\vec{R}_n + \vec{R}_m - \vec{R}_{n'} - \vec{R}_{m'})/2] \tag{8.34}
$$

as a convolution integral

$$
\tilde{c}_{ij}^{(e)}(\lambda) = n_b/(2\pi)^3 \int d^3k \, S_b(k)S_{ij}(\lambda,\vec{q}_\lambda - \vec{k}) . \tag{8.35}
$$

As an illustration of the possible form of the resulting coupling function $\tilde{c}_{ij}^{(e)}(\lambda)$, consider the case of a Gaussian decay of the conditional correlation function $S_{ij}(\lambda,\vec{r})$. This case leads to the following wave-vector dependence of the coupling:

$$
\tilde{c}_{ij}^{(e)}(\lambda) \propto \overline{\sin^2(\vec{q}_\lambda \cdot \vec{R}_{nm}/2)} \cdot \bar{S}_b(q_\lambda) , \tag{8.36}
$$

where the first factor arises from the average of $\alpha_i^{(nm)}(\lambda)\left(\alpha_j^{(nm)}(\lambda)\right)^*$ taken over all bonds (nm), and $\bar{S}_b(q_\lambda)$ is the smeared-out bond structure factor. For a random distribution of the bond centers, $S_b(q)$ equals unity. In this case one obtains the approximate result

$$
\tilde{c}_{ij}^{(e)}(\lambda) \propto 1 - \sin(q_\lambda d)/(q_\lambda d) , \tag{8.37}
$$

where d is the bond length. It is interesting that this result, in contrast to the Martin and Brenig ansatz (8.20), does not contain a correlation length related to the range of atomic order. Although the result (8.37) is derived under simplifying assumptions, it indicates again the difficulties involved in a correct structural interpretation of low-frequency Raman data.

8.3 Quasielastic Spectrum

8.3.1 Experimental Results

Although optical glasses possess a high degree of homogeneity on the scale of optical wavelengths, the total intensity of scattered light is dominated by the elastic component caused by the existing optical inhomogeneity. For example, in vitreous silica at room temperature, the scattered intensity in one Brillouin line amounts only to 2% of the intensity of the elastically scattered light [8.25]. For silica-based glasses this Rayleigh line is only weakly depolarized [8.26]. This result indicates that the optical inhomogeneities are very nearly isotropic and can be attributed to the static density fluctuations frozen-in at the glass tran-sition temperature [8.27]. The structural origin of the isotropy is the cubic sym-metry of the tetrahedral SiO_4 units of vitreous silica. In other glasses like a-B_2O_3 [8.26], where the basic structural unit lacks cubic symmetry, the static optical inhomogeneities are to a large degree anisotropic, and the elastically scattered light is strongly depolarized.

The high intensity of the strictly elastic component, compared with the inten-sity of the inelastically scattered light, presents a serious problem to the in-vestigation of any quasielastic scattering because of the finite width of the in-strumental profile. Another difficulty in defining and resolving the quasielastic component is the overlap with the first-order vibrational contribution which ex-tends down to zero, as described above. Using double-grating spectrometers WINTER-LING [8.13] and NEMANICH [8.12] were able to resolve the inelastic scattering down to 3 cm^{-1}. FLEURY and LYONS [8.28] and HEIMAN et al. [8.20] used an iodine absorp-tion cell to eliminate the elastically scattered light completely and a double-(triple)-pass Fabry-Pérot interferometer for the spectroscopy of the inelastically scattered light. In this way FLEURY and LYONS could resolve the quasielastic line of LaSF-7 glass to within about 1 GHz of the line center.

A quasielastic component in the light scattering spectrum of glasses was first discovered for vitreous silica by WINTERLING [8.13] and for B_2O_3 and As_2S_3 glasses by WINTERLING [8.29] and WINTERLING and ARAI [8.30]. For the reason mentioned above they did not obtain a complete picture of a quasielastic line, but they ob-served an upturn of the spectral intensity towards lower frequencies in the range 5 - 10 cm^{-1} at temperatures from 80 K to room temperature. At $|\bar{\omega}| = 5$ cm^{-1} the

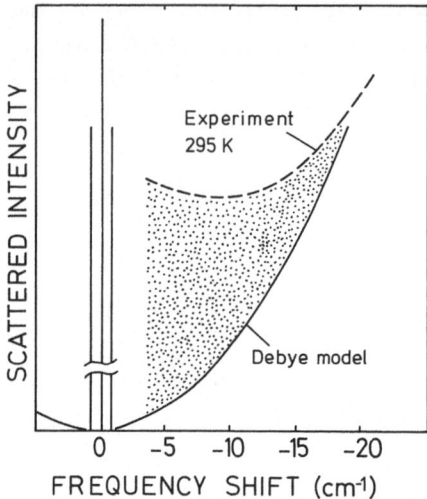

Fig.8.6. Light scattering spectrum of vitreous silica at low frequencies (schematic). The inner three lines represent the Rayleigh line and the Brillouin doublet [8.13]

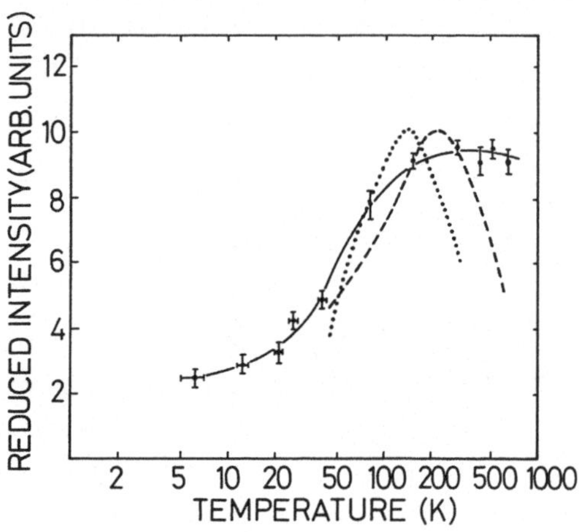

Fig.8.7. Temperature dependence of the reduced depolarized Raman intensity of vitreous silica (Suprasil W1) at $\Delta\omega = 5$ cm^{-1} [8.30]. Dashed line: Theory [8.18], dotted line: Brillouin line width at 33 GHz [8.38]

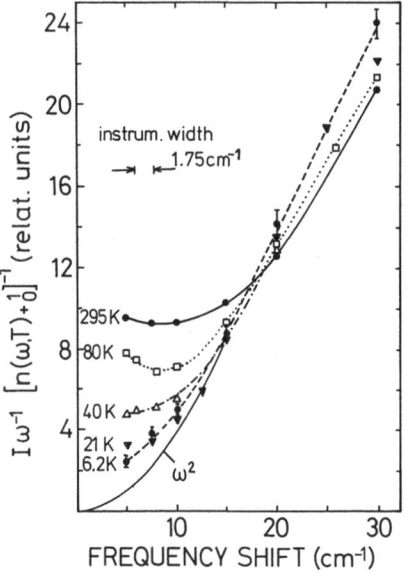

Fig.8.8. Reduced spectra of the depolarized scattering in vitreous silica (Suprasil W1) at various temperatures. The ω^2-curve indicates the limiting behavior of the first-order vibrational spectrum [8.30]

measured intensity at room temperature was about two orders of magnitude higher than the intensity in the instrumental tail of the elastic component. The first-order vibrational contribution, on the other hand, can be estimated to amount to less than 25% of the total density at this frequency and temperature (see Fig.8.8). The different parts of the light scattering spectrum are drawn schematically in

Fig.8.6. The "light scattering excess" between the extrapolated curve for the first-order vibrational contribution and the measured intensity represents the wing of a quasielastic component. The excess scattering exhibits a characteristic temperature dependence, as shown in Figs.8.7 and 8.8. Contrary to the first-order vibrational contribution above about 20 cm^{-1}, the reduced intensity $I^R(\omega)$ in the frequency range 5 - 15 cm^{-1} increases rather strongly with increasing temperature up to room temperature. The temperature dependence is, however, weaker than for a second-order difference process with harmonic vibrational modes. The difference in temperature dependence can be used to separate the quasielastic from the first-order vibrational component in the experimental Raman spectrum [8.12]. Such an analysis has the advantage of being independent of the extrapolation of the vibrational contribution to low frequencies, which relies on the validity of the quadratic frequency dependence of the Raman coupling $C_{ij}(\omega)$ and of the Debye density of states $g(\omega)$. In contrast to the temperature dependence, the depolarization ratio for the quasielastic component is practically identical to that for the first-order vibrational contribution in the frequency region below the "boson peak". WINTERLING [8.29] reported depolarization ratios I_{HV}/I_{VV} of 0.30 ± 0.03 for vitreous silica, of 0.47 ± 0.05 for a-B_2O_3, and of 0.45 ± 0.06 for a-As_2S_3.

NEMANICH [8.12] measured the low-frequency Raman spectra of the chalcogenide glasses As_2S_3, As_2Se_3, GeS_2, and $GeSe_2$, and of the alloy system $(As_2S_3)_{1-x}(GeS_2)_x$. He found that the quasielastic contribution exists in all these materials, but is only strong enough for a quantitative analysis in a-As_2S_3 and the alloy system. NEMANICH observed no upturn at the lower end of the spectrum of the inelastically scattered light, but the temperature dependence of the light scattering excess is qualitatively similar to Winterling's results. The experimental data actually show an upturn of the scattered intensity at the lowest frequencies $|\bar{\omega}| \leq 4$ cm^{-1}, but this feature can be ascribed to the instrumental tail of the elastic Rayleigh line according to the measured depolarization ratio of 0.25 which agrees with the value for the elastic component in a-As_2S_3. For the frequency range $5 \leq |\bar{\omega}| \leq 50$ cm^{-1} NEMANICH reported nearly constant depolarization ratios I_{HV}/I_{VV} of about 0.45, 0.5, and 0.43 for a-As_2S_3, a-GeS_2, and a-$GeSe_2$, respectively.

FIRSTEIN et al. [8.31] first used the iodine filter in measuring the low-frequency Raman spectrum of a glass and found a quasielastic component in the Schott glass LaSF-7. This glass, which contains mostly B_2O_3, La_2O_3, and ThO_2 with a few percent of Ta_2O_5 and Nb_2O_5, is one of the strongest Raman scatterers among the optical glasses. FLEURY and LYONS [8.28a] improved Firstein's method by using a Fabry-Pérot interferometer. Correcting the measured spectrum for the iodine absorption spectrum, they obtained the quasielastic component in the LaSF-7 glass as a Lorentzian with a half-width (HWHM) of 0.12 cm^{-1} at room temperature (Fig. 8.9). The half-width increases by a factor of 8 when the temperature is raised from 100 to 400 K. FLEURY and LYONS [8.28b] detected a narrow quasielastic component

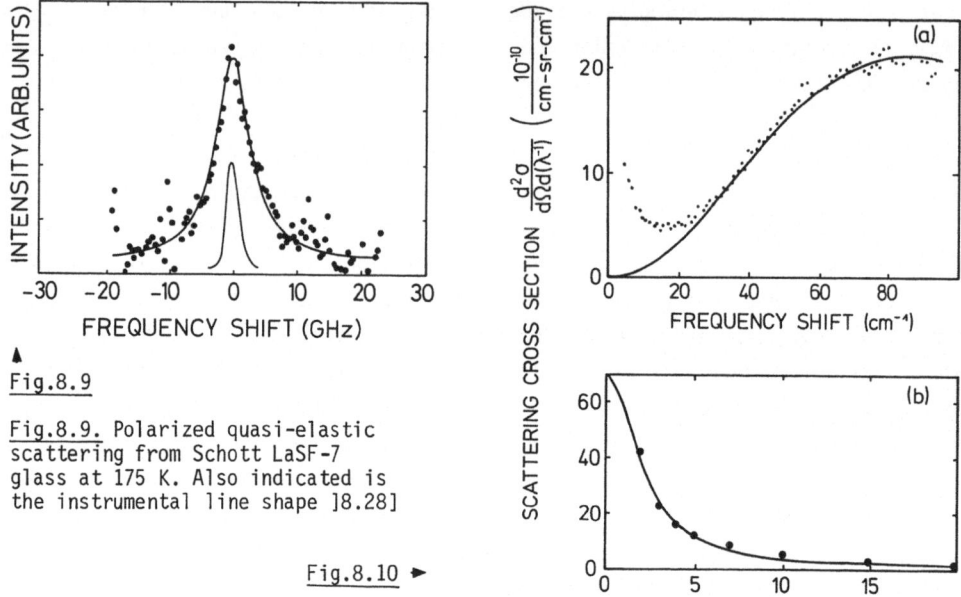

Fig.8.9

Fig.8.9. Polarized quasi-elastic
scattering from Schott LaSF-7
glass at 175 K. Also indicated is
the instrumental line shape]8.28]

Fig.8.10 ➤

Fig.8.10. Low-frequency Raman scattering from LaSF-7 glass in absolute units [8.20].
a) First-order vibrational spectrum and light scattering excess at room temperature.
The solid line is a fit of the Martin-Brenig theory to the data. b) The light scattering excess. The solid line is a Lorentzian fit having a half-width (HWHM) of
2.5 cm^{-1}

also in seven other glasses composed mainly of transition-metal or rare earth oxides,
but not in vitreous silica and in other optical glasses. In a recent paper, HEIMAN
et al. [8.20] reported the observation of the broad quasielastic component in fifteen
optical glasses, including vitreous silica and the LaSF-7 glass. For every glass
the quasielastic intensity scales roughly with the total Raman scattering intensity.
It is, therefore, strongest in the LaSF-7 glass. For this material the excess of the
scattering intensity over the extrapolated first-order vibrational contribution can
be fitted by a Lorentzian with a half-width (HWHM) of (2.5 ± 1) cm^{-1} (Fig.8.10).
Combined with Fleury and Lyons' observation that the quasielastic line between
$\bar{\omega} = 0$ and $|\bar{\omega}| \approx 0.7$ cm^{-1} is a Lorentzian with a half-width of 0.12 cm^{-1}, this result implies that the total quasielastic spectrum of LaSF-7 glass is non-Lorentzian,
but can be decomposed into a narrow and a broad Lorentzian component.

Concerning the possible explanation of these data for glasses, some information
about quasielastic light scattering in crystals may be useful. LYONS and FLEURY
[8.32] observed a quasielastic spectrum of crystalline $KTaO_3$ which consists, at a
temperature of 182 K, of a narrow line with a half-width of 2.5 GHz superimposed on
a broader line of 40 GHz half-width. The narrow line is identified as arising from
the heat conduction mode, and the broad component is tentatively ascribed to two-
phonon difference processes or phonon density fluctuations [8.33]. Quasielastic
light scattering has also been found in superionic conductors. WINTERLING et al.

[8.34] reported the observation of a relatively narrow Lorentzian line of about 4 cm^{-1} half-width superimposed on a broad quasielastic spectrum for α-AgI at 178° C. The narrow line can be shown to arise from the hopping motion of the mobile Ag$^+$ ions in the crystalline lattice of the I$^-$ ions. (For a recent theoretical review on superionic conductors see [8.35].

8.3.2 Theory

a) *Physical Origin of the Quasielastic Scattering*

Before considering scattering mechanisms which are specific to glass, it is reasonable first to examine contributions to the quasielastic scattering which also exist in the crystalline case; such are the contributions from heat conduction, ionic hopping and two-phonon difference processes. Since the thermal conductivity κ of glasses at room temperature is very low, the contribution of the heat conduction mode in glasses has a negligible width compared with experimental values. For vitreous silica at room temperature the half-width $\Gamma = D_T k^2$ in terms of the thermal diffusivity $D_T = \kappa/C_p$, where C_p is the specific heat at constant pressure, is only of the order of 10^{-3} cm^{-1}. The hopping frequency of ionic network modifiers in glasses at room temperature is also much too low in comparison with observed quasielastic line widths. The two-phonon difference processes with the normal vibrations in glass lead to a scattering intensity given by

$$I(\omega) = \sum_{\lambda_1,\lambda_2} |\chi(\lambda_1,\lambda_2)|^2 n[\omega(\lambda_1)]\{1 + n[\omega(\lambda_2)]\}$$

$$\frac{[\Gamma(\lambda_1) + \Gamma(\lambda_2)]/\pi}{[\omega(\lambda_1) - \omega(\lambda_2) + \omega]^2 + [\Gamma(\lambda_1) + \Gamma(\lambda_2)]^2} \quad , \tag{8.38}$$

where the Lorentzian results from the finite lifetimes $1/[2\Gamma(\lambda)]$ of the normal modes λ, which are due to the anharmonic mode interactions. If the second-order polarizability $\chi(\lambda_1,\lambda_2)$ is a smoothly varying function of λ_1 and λ_2, this lifetime broadening has little effect and can be neglected. In this case the spectrum (8.38) has a width of the order of $k_B T$, and the intensity for $\omega > 0$ increases at least quadratically with temperature, both of which are in disagreement with experiment. If, on the other hand, the coupling parameter $\chi(\lambda_1,\lambda_2)$ is diagonal, viz.,

$$\chi(\lambda_1,\lambda_2) = \chi^{(2)}(\lambda_1)\delta_{\lambda_1,\lambda_2} \quad , \tag{8.39}$$

the expression (8.38) yields

$$I(\omega) = \sum_{\lambda} |\chi^{(2)}(\lambda)|^2 n[\omega(\lambda)]\{1 + n[\omega(\lambda)]\}\frac{2\Gamma(\lambda)/\pi}{\omega^2 + [2\Gamma(\lambda)]^2} \tag{8.40}$$

and describes the contribution of occupation number fluctuations of the vibration modes to the scattering spectrum. Assuming that

$$\chi^{(2)}(\lambda) = A\omega(\lambda) \quad (A = \text{const.}) \tag{8.41}$$

and replacing the mode widths $\Gamma(\lambda)$ by their thermal average $\bar{\Gamma}$, one obtains the following result:

$$I(\omega) = A^2 k_B^2 T^2 C_v \frac{2\bar{\Gamma}/\pi}{\omega^2 + (2\bar{\Gamma})^2} \quad . \tag{8.42}$$

Here C_v is the vibrational specific heat at constant volume. The analogue of this formula for the ultrasonic absorption has been used by PINE [8.36] to explain the damping of ultrasonic waves in vitreous silica (see below). The average width $\bar{\Gamma}$ in (8.42) can be estimated from the thermal conductivity of the crystalline form of the material if it is assumed that the rate of anharmonic processes is similar in glass and crystal. Using a Debye phonon model for quartz at room temperature, a half-width $2\bar{\Gamma}$ of the order of 5 cm^{-1} is obtained. Since the average width $\bar{\Gamma}$ due to anharmonic interactions increases monotonously with increasing temperature, (8.42) predicts a very strong temperature dependence in the wing $\omega > 2\bar{\Gamma}$ of the quasielastic component. This result is not compatible with the experimental data for vitreous silica and the chalcogenide glasses at room temperature and lower temperatures. Therefore, for both forms of the second-order vibrational spectrum given by (8.38,42), the temperature dependence of the scattered intensity is too rapid when compared with available experimental data. Consequently, two-phonon difference processes are unlikely to explain the observed quasielastic spectra of glasses.

A fourth case to be examined is the possibility that the light scattering excess represents a broad tail of the Brillouin spectrum rather than the wing of a quasielastic component centered at zero frequency. For the frequency range 2 - 20 cm^{-1}, which is well above the Brillouin lines, such an explanation can be ruled out in the following way. A tail in the Brillouin spectrum results from a frequency-dependent contribution $\delta C(\omega)$ to the elastic constant, which characterizes the anelastic properties of the medium. For the Maxwell model of stress relaxation, in particular, $\delta C(\omega)$ is given by

$$\delta C(\omega) = -\frac{\Delta C}{1 - i\omega\tau} \quad , \quad \Delta C > 0 \quad , \tag{8.43}$$

with a relaxation time τ. Expanding the elastic response function for $\omega \gg ck$ to lowest order in $\delta C(\omega)$, one obtains a contribution to the scattered intensity proportional to

$$\text{Im}\left\{\frac{1}{-\omega^2 + k^2[c^2 + \delta C(\omega)/\rho]}\right\} \approx -(k^2/\rho)\,\text{Im}[\delta C(\omega)/\omega^4] \tag{8.44}$$

in the frequency region considered. Independent of the particular form of the an-elastic part $\delta C(\omega)$ of the elastic constant, the factor ω^{-4} leads to a very rapid decay of this contribution with increasing frequency, which is incompatible with the experimental data. This conclusion is confirmed by the observation of different depolarization ratios for the Brillouin lines and for the excess scattering.

After examination of scattering mechanisms which also occur in crystalline material, contributions to the quasielastic spectrum which are specific to glasses now need to be considered. The first of these is an extension of (8.44) in the absence of translational symmetry. It is analogous to the substitution of the con-tinuous first-order vibrational spectrum of glasses for the discrete Brillouin-Raman spectrum of crystals as described in Section 8.2 and was first discussed by WINTERLING [8.13]. The deviation of the glass network from strictly harmonic be-havior can be described by replacing the square $\omega^2(\lambda)$ of the harmonic eigenfre-quency by $[\omega^2(\lambda) + M(\lambda,\omega)]$, in the response function for the vibration mode λ. $M(\lambda,\omega)$ may be called the "memory function". A possible choice of this memory func-tion is an expression analogous to (8.43) for the Maxwell relaxation model. A simi-lar assumption is the basis of FULDE and WAGNER's [8.37] theory of the thermal properties of glasses at low temperatures. Expanding the response function for the vibration mode for $\omega \ll \omega(\lambda)$ to lowest order in $M(\lambda,\omega)$, a quasielastic contribution to the Raman spectrum proportional to the following sum over all vibration modes is obtained:

$$\text{Im}\left\{\sum_\lambda \frac{\tilde{C}_{ij}(\lambda)}{-\omega^2 + \omega^2(\lambda) + M(\lambda,\omega)}\right\} \approx -\sum_\lambda [\tilde{C}_{ij}(\lambda)/\omega^4(\lambda)]\text{Im}\{M(\lambda,\omega)\} \quad . \qquad (8.45)$$

Here $\tilde{C}_{ij}(\lambda)$ is again the coupling of vibration mode λ to Raman scattering (8.5). Because of the factor $\omega^{-4}(\lambda)$ on the right hand side of (8.45), one may expect that the modes with lower frequency give the largest contribution to this expression. This would explain why the measured depolarization ratios of the light scattering excess and the first-order vibrational spectrum at lower frequencies up to the boson peak are nearly the same. Equation (8.45) provides a formal framework for the description of quasielastic scattering in glasses. The atomistic origin of the memory function in (8.45) is not specified in this description. Therefore, a model is needed which contains more specific assumptions and thereby explains the structural origin of the relaxation behavior expressed by the memory function. The "defect model" provides such an explanation. This model describes a set of struc-tural defects in glass, each of which can exist in two different atomic confi-gurations ("states") and make thermally activated transitions between them. Since the defect model was originally proposed as an explanation of ultrasonic absorption experiments, it will first be shown how the low-frequency Raman scattering can be related to the ultrasonic absorption.

b) *Relation Between Raman Scattering and Ultrasonic Absorption*

The Raman tensor which determines the intensity of inelastically scattered light
is given by the correlation function of the space- and time-dependent fluctuations
of the tensor components x_i of the dielectric susceptibility (8.1),

$$I_{ij}(\omega) = (1/2\pi) \int_{-\infty}^{+\infty} dt\, e^{i\omega t} \langle x_i(\vec{k},t) x_j(-\vec{k},0)\rangle \quad . \tag{8.46}$$

Here, $x_i(\vec{k})$ is the spatial Fourier transform and $\vec{k}$ is the wave-vector transfer in
the scattering. As a consequence of the fluctuation-dissipation theorem, the absorption of an elastic wave of polarization i (i = l or t for longitudinal or
transverse polarization) is determined by the correlation function for the fluctuations of the corresponding component e_i of elastic strain

$$S_{e_i,e_i}(k,\omega) = (1/2\pi) \int_{-\infty}^{+\infty} dt\, e^{i\omega t} \langle e_i(\vec{k},t) e_i(-\vec{k},0)\rangle \quad . \tag{8.47}$$

The absorption coefficient $\alpha_i(\omega)$, defined by the exponential decay of the intensity of the propagating wave with distance, is given by

$$\alpha_i(\omega) = a_i \omega S_{e_i,e_i}(\omega/c_i,\omega)/[1 + n(\omega)] \tag{8.48}$$

with

$$a_i = \rho_0 c_i/2\hbar \quad , \tag{8.49}$$

where ρ_0 is the mass density of the glass and c_i the sound velocity for polarization
i. The factor $[1 + n(\omega)]$ can usually be replaced by $k_B T/(\hbar\omega)$. If the inelastic
scattering and the absorption have the same physical origin, the fluctuations (8.46)
of the dielectric susceptibility and the fluctuations (8.47) of the elastic strain
are caused by the time-dependent fluctuations of the same internal degree of freedom of the medium. In this case the following relation between the Raman tensor
$I_{ij}(\omega)$ and the ultrasonic absorption coefficient $\alpha_i(\omega)$ holds:

$$I_{ij}(\omega) = a'_{ijk}[1 + n(\omega)]\alpha_k(\omega)/\omega \quad , \tag{8.50}$$

with constant coefficients a'_{ijk}. Conversely, by examining the validity of this relation it can be tested whether Raman scattering and ultrasonic absorption have
the same physical origin. This test is possible before a particular physical
mechanism is specified. Figure 8.7 shows the temperature dependence of the reduced depolarized Raman intensity at 5 cm^{-1} together with the ultrasonic absorption
at 33 GHz ($\hat{=}$ 1 cm^{-1}) as deduced from the width of the longitudinal Brillouin line
for vitreous silica. Both curves show a similar increase between about 50 and
150 K, but the behavior at higher temperatures is different. In contrast to the
Raman data, the ultrasonic absorption curve exhibits a pronounced maximum. For a
phonon frequency corresponding to 5 cm^{-1} the position of the absorption maximum

would be close to room temperature. It may, therefore, be concluded that the physical origin of the Raman spectrum and the ultrasonic absorption is probably the same below room temperature, but that at higher temperatures a second mechanism contributes exclusively to the light scattering.

The apparent relationship between the quasielastic Raman scattering and the ultrasonic absorption below room temperature lends further support to the conclusion that the Raman data for glass cannot be explained by two-phonon difference processes. Although PINE [8.36] has shown that the ultrasonic absorption in fused silica can be fitted to the anharmonic phonon model [corresponding to (8.42) for the Raman scattering] with fewer parameters than in the case of the structural defect model, the experimental facts point to a mechanism which is specific to the glass structure. The first point of this argument is that the ultrasonic attenuation in vitreous silica below room temperature is not only much stronger than in crystalline quartz, but shows a rather different temperature dependence (see Chap.6). Whereas for quartz the part of the absorption curve following the rapid increase up to about 50 K is flat, a rather sharp peak is observed for vitreous silica. Secondly, this peak is more specific to the glass material than can be expected for any anharmonic phonon process. Von HAUMEDER et al. [8.39] observed that for SiO_x films ($0 \leq x \leq 2$) the absorption peak of pure vitreous silica (at $T \approx 120$ K for 33 GHz) becomes weaker and finally disappears with decreasing oxygen content, but stays at the same temperature for a given frequency. Such a behavior has never been seen in the ultrasonic absorption caused by anharmonic phonons in crystals. Von Haumeder's result confirms the original suggestion of ANDERSON and BÖMMEL [8.40] that the ultrasonic loss peak in vitreous silica is related to the flexibility of the oxygen bridges connecting the tetrahedrally coordinated silicon atoms. More specifically, ANDERSON and BÖMMEL suggested that, for a certain fraction of oxygen bridges in the glass network, two stable configurations exist. Time-dependent fluctuations causing absorption and scattering are induced by thermally activated transitions between the two configurations. In essence, such oxygen bridges represent localized defects with an internal degree of freedom corresponding to the two different configurations. A defect model of this sort can be formulated under more general conditions if it is assumed that thermally activated transitions between different structural configurations occur locally in certain nonoverlapping regions of limited size (not more than 10 - 20 Å in diameter, say). Such an assumption is the basis of the defect model to be described in more detail in the following section.

c) *Defect Model*

In order to fit the observed temperature dependence of the ultrasonic absorption in vitreous silica, ANDERSON and BÖMMEL [8.40] assumed that different groups of defects with a broad distribution of relaxation times exist (see Chap.6). More

recently, HUNKLINGER [8.41] obtained satisfactory fits of the temperature dependence of the ultrasonic absorption in vitreous silica for various frequencies
using a distribution of activation energies

$$P(V) = \bar{P} \exp[- (V - V_m)^2/V_0^2 - C^2/V^2]$$ (8.51)

with parameters $\bar{P} = 5.6 \times 10^{17}$ K^{-1} cm^{-3}, $V_m/k_B = 410$ K, $V_0/k_B = 550$ K and
$C/k_B = 125$ K. The distribution of relaxation times is related to P(V) by the
Arrhenius formula

$$\tau(V) = \tau_0 \exp(V/k_B T)$$ (8.52)

where a value $\tau_0 = 2 \times 10^{13}$ s was used. THEODORAKOPOULOS and JÄCKLE [8.18] assumed
that the hopping of a defect between its two possible configurations contributes
to the inelastic scattering of light, since the dielectric susceptibility of a
defect is different in its two configurations by an amount $\Delta\alpha$. The degree of depolarization of the scattered light depends on the form of this difference tensor.
If it has only one nonzero eigenvalue $\Delta\alpha^{(3)}$ one obtains a depolarization ratio

$$I_{VH}/I_{VV} = 1/3$$ (8.53)

and a polarized Raman intensity proportional to

$$I_{VV}(\omega) \propto [\Delta\alpha^{(3)}]^2 \omega[1 + n(\omega)] \int dVdE \; P(V,E) \left[- \frac{df(E)}{dE}\right] \frac{\tau(V)}{1 + \omega^2\tau^2(V)} \; .$$ (8.54)

Here $n(\omega)$ is the Bose function, $f(E) = [\exp(E/k_B T) + 1]^{-1}$, and P(V,E) is the joint
distribution function of the activation energy V and the free-energy difference E
which exists between the two defect configurations. The same authors calculated
the temperature dependence of the reduced Raman intensity for a joint distribution
function

$$P(V,E) = P(V) \cdot \delta(E)$$ (8.55)

using a barrier distribution P(V) similar to Hunklinger's equation (8.51). The
result of the calculation for $|\bar{\omega}| = 5$ cm^{-1} is shown by the dotted line in Fig.8.7.
The calculated temperature dependence agrees reasonably well with the experimental
curve for vitreous silica at temperatures not higher than room temperature, but
fails to agree at higher temperatures. The similarity of the calculated reduced
Raman intensity with the temperature dependence of the ultrasonic absorption
shown in the same figure makes it clear that this discrepancy is not due to an in-
accuracy of the distribution function P(V,E), but indicates a different mechanism
for the Raman scattering above room temperature. The nature of this additional
mechanism is not clear at present. As for the frequency dependence, the room-
temperature Raman spectrum obtained with Hunklinger's P(V) distribution (8.51)
is strongly non-Lorentzian, and can approximately be decomposed into two Lorentzians

with half-widths (HWHM) of 0.5 and 5 cm^{-1}. The latter value is in reasonable agreement with a width $\Gamma = 8 \pm 3$ cm^{-1} determined by HEIMAN et al. [8.20] from a Lorentzian fit to Raman data for vitreous silica in the frequency region 5 - 30 cm^{-1}. NEMANICH [8.12] has tested the validity of (8.54) for the light scattering excess in the chalcogenide glass As_2S_3 using information about the barrier distribution P(V) from ultrasonic absorption measurements [8.42]. He was not able to fit both the frequency and the temperature dependence satisfactorily with a single set of parameters. Unfortunately, Brillouin scattering data for a-As_2S_3 are not available so that a direct comparison of ultrasonic absorption and quasielastic light scattering, as in Fig.8.7 for vitreous silica, cannot be made. It does not seem impossible, therefore, that the discrepancy stated by NEMANICH is due to an inaccurate choice of the distribution function P(V,E) rather than to an inadequacy of the model.

For the Schott glass LaSF-7 and a few similar glasses the quasielastic scattering has been determined both below and above the longitudinal Brillouin line [8.28, 20], respectively. Although this glass is distinguished by its strong Raman activity, which is due to the large content of heavy metal ions, possibly the quasielastic scattering is caused by a similar kind of structural relaxation as assumed for the other glasses. If the distribution P(V) of activation energies is suitably chosen, the defect model can of course reproduce the observed quasielastic spectrum of LaSF-7 glass which consists of a narrow and a broad Lorentzian component. It needs to be checked, however, whether such a distribution is consistent with ultrasonic properties. It would also be interesting to measure the quasielastic scattering below the Brillouin spectrum for the more common glasses like vitreous silica.

The depolarization ratio (8.53), obtained for a uniaxial defect polarizability, agrees within experimental error with the measured value for vitreous silica (0.30 ± 0.03). To fit the absolute magnitude of the light scattering excess, a value of $\Delta\alpha^{(3)} = 0.6$ $\mathring{A}^3$ is required for vitreous silica [8.18]. This is not an unreasonable value compared with the polarizability of 1.7 $\mathring{A}^3$ of O^{--} in crystalline SiO_2 [8.43]. Different values of the depolarization ratio, as observed for other glasses, can be obtained for more complicated forms of the polarizability difference tensor $\Delta\alpha$. The model can in fact yield any value between 0 and 3/4 for a suitable choice of the three eigenvalues $\Delta\alpha^{(\lambda)}$ of the tensor $\Delta\alpha$ according to the general formula [8.44]

$$I_{VH}/I_{VV} = (1/2) \frac{\gamma - 1}{1 + 2\gamma/3} \tag{8.56}$$

where γ is defined by

$$\gamma = 3 \sum_{\lambda=1}^{3} \left(\Delta\alpha^{(\lambda)}\right)^2 / \left(\sum_{\lambda=1}^{3} \Delta\alpha^{(\lambda)}\right)^2 . \tag{8.57}$$

In the version of the defect model described so far, it is assumed that the hopping of the defects is directly coupled to the scattered light through the polarizability difference between the two defect configurations. The form of the difference tensor appears as characteristic of the defects in a particular glass structure, but the observed agreement between the depolarization ratios of the light scattering excess and the first-order vibrational spectrum seems fortuitous and remains unexplained. However, the results derived from the defect model can also be interpreted in terms of "indirect coupling" whereby the defects couple to the light only via the vibrational degrees of freedom. In this way the agreement between the measured depolarization ratios for the two parts of the Raman spectrum can be made plausible. The interpretation in terms of indirect coupling can be justified using the general expression (8.45) which is based on the FULDE and WAGNER theory [8.37]. In fact, if the mode memory function $M(\lambda,\omega)$ occuring in this expression is derived from the defect model, a result for the light scattering excess is obtained which is formally identical with (8.54); the polarization difference $\Delta\alpha$ of the defects is replaced by the product of the coupling $\tilde{C}_{ij}(\lambda)$ between mode λ and the light and the coupling of the same mode to the configuration fluctuations of the defects. In physical terms this replacement corresponds to attributing the polarizability difference associated with a defect to the elastic polarization of the glass network surrounding the defect, rather than to the defect center.

As a further consequence of the defect model, THEODORAKOPOULOS and JÄCKLE predicted an observable contribution to the low-frequency Raman spectrum from tunneling defects in glasses as postulated by ANDERSON et al. [3.45] and PHILLIPS [8.46]. So far, such a contribution has not been detected experimentally.

8.4 Conclusion

It follows from the interpretation of the experimental results given above that both contributions to the low-frequency Raman scattering in glasses arise from the disordered atomic structure. As for the first-order vibrational spectrum (Sect.8.2), the effect of the disordered structure has been analysed in terms of an electrical and a mechanical disorder. The two types of disorder are characterized by correlation functions of elastooptical and vibrational properties, and the same quadratic frequency dependence is obtained for both contributions to the Raman coupling at low frequency. It has been shown that for vitreous silica the electrical disorder gives the major contribution at low frequencies. In the past, attempts have been made to extract structural information from low-frequency Raman data in terms of a structural correlation length. However, in order to obtain meaningful results, the effect of the precise frequency dependence of the vibrational density of states on the Raman intensity needs first to be subtracted. But even at this point the correct structural interpretation of the Raman coupling is a nontrivial problem, since dispersion effects due to the discrete atomic structure occur even for complete randomness. This has been illustrated using the bond polarizability model.

Since the physical mechanism leading to the quasielastic scattering in glasses (Sect.8.3) is not immediately obvious, several possible causes, including two phonon-difference processes, have been discussed first. It has been shown that both the frequency and temperature dependence of the spectrum and the comparison with ultrasonic data support the idea that the quasielastic scattering is due to a mechanism specific to the glass structure. Following the interpretation of ultrasonic absorption in glasses below room temperature, this mechanism is described by a hopping model for localized structural defects with two atomic configurations. In vitreous silica at temperatures above room temperature, however, the different temperature dependence of the quasielastic scattering and ultrasonic absorption points to an additional contribution, the origin of which remains unclear. Another point which needs to be tested is whether the explanation by the defect model also applies to the inner part of the quasielastic spectrum in the frequency region between zero and the longitudinal Brillouin line. In this frequency range, so far, the quasielastic scattering has been investigated only for the strongly scattering Schott LaSF-7 glass and other transition-metal and rare-earth glasses, where a comparatively narrow line has been observed. Clearly, a further experimental investigation of the quasielastic scattering both at very low frequencies and at high temperatures is desirable. To clarify the connection with the ultrasonic absorption, a combination with measurements of the Brillouin line width would be ideal.

Acknowledgements. I wish to thank Drs. G. Winterling and J.R. Sandercock for useful information on the subject matter of this article.

References

8.1   M.H. Brodsky: In *Light Scattering in Solids*, Topics in Applied Physics, Vol.8, ed. by M. Cardona (Springer, Berlin, Heidelberg, New York 1975) p.205
8.2   G. Lucovsky: In *Amorphous and Liquid Semiconductors*, ed. by J. Stuke, W. Brenig (Taylor and Francis, London 1974) p.1099
8.3a R. Shuker, R.W. Gamon: Phys. Rev. Lett. *25*, 222 (1970)
8.3b R. Shuker, R.W. Gamon: In *Light Scattering in Solids* (Flammarion Sciences, Paris 1971) p.334
8.4   E. Whalley, J.E. Bertie: J. Chem. Phys. *46*, 1264 (1967)
8.5   R.H. Stolen: Phys. Chem. Glasses *11*, 83 (1970)
8.6   P. Flubacher, A.J. Leadbetter, J.A. Morrison, B.P. Stoicheff: J. Phys. Chem. Solids *12*, 53 (1959)
8.7   A.J. Martin, W. Brenig: Phys. Status Solidi b *64*, 163 (1974)
8.8   J.E. Smith, Jr., M.H. Brodsky, B.L. Crowder, M.I. Nathan, A. Pinczuk: Phys. Rev. Lett. *26*, 642 (1971)
8.9   F.L. Galeener, G. Lucovsky: In *Light Scattering in Solids*, ed. by M. Balkanski, R.C.C. Leite, S.P.S. Porto (Flammarion Sciences, Paris 1976) p.641
8.10 R.J. Bell, N.F. Bird, P. Dean: J. Phys. C *1*, 299 (1968)
8.11 J.S. Lannin: Solid State Commun. *12*, 947 (1973)
8.12 R.J. Nemanich: Phys. Rev. B*16*, 1655 (1977)
8.13 G. Winterling: Phys. Rev. B*12*, 2432 (1975)

8.14 W. Prettl, N.J. Shevchik, M. Cardona: Phys Status Solidi b *59*, 241 (1973)
8.15 J. Jäckle, K. Froböse: J. Phys. F *9*, 967 (1979)
8.16 J. Jäckle: In *The Physics of Non-Crystalline Solids*, ed. by G.H. Frischat (Trans Tech. Publ., Aedermannsdorf 1977) p.568
8.17 H.E. Shull, K. Vedam: J. Appl. Phys. *43*, 3724 (1972)
8.18 N. Theodorakopoulos, J. Jäckle: Phys. Rev. B*14*, 2637 (1976)
8.19 R.J. Nemanich, M. Gorman, S.A. Solin: Solid State Commun. *21*, 277 (1977)
8.20 D. Heiman, R.W. Hellwarth, D.S. Hamilton: J. Non Cryst. Solids *34*, 63 (1979)
8.21 R. Alben, D. Weaire, J.E. Smith, Jr., M.H. Brodsky: Phys. Rev. B*11*, 2271 (1975)
8.22 R.J. Bell, D.C. Hibbins-Butler: J. Phys. C *9*, 2955 (1976)
8.23 D. Beeman, R. Alben: Adv. Phys. *26*, 339 (1977)
8.24 R.J. Bell, A. Carnevale, C.R. Kurkjian, G.E. Peterson: J. Non Cryst. Solids *35/36*, 1185 (1980)
8.25 T.C. Rich, D.A. Pinnow: Appl. Phys. Lett. *20*, 264 (1972)
8.26 J.A. Bucaro, H.D. Dardy: J. Appl. Phys. *45*, 2121 (1974)
8.27 N.L. Laberge, V.V. Vasilescu, C.J. Montrose, P.B. Macedo: J. Am. Ceram. Soc. *56*, 506 (1973)
8.28a P.A. Fleury, K.B. Lyons: Phys. Rev. Lett. *36*, 1188 (1976)
8.28b P.A. Fleury, K.B. Lyons: In *Structure and Excitations of Amorphous Solids*, Conference Proceedings No.31, ed. by G. Lucovsky, F.L. Galeener (American Institute of Physics, New York 1976) p.263
8.29 G. Winterling: In *Light Scattering in Solids*, ed. by M. Balkanski, R.C.C. Leite, S.P.S. Porto (Flammarion Sciences, Paris 1976) p.663
8.30 G. Winterling, T. Arai: In *The Physics of Non-Crystalline Solids*, ed. by G.H. Frischat (Trans Tech. Publ., Aedermannsdorf 1977) p.580
8.31 L.A. Firstein, J.M. Cherlow, R.W. Hellwarth: Appl. Phys. Lett. *28*, 25 (1976)
8.32 K.B. Lyons, P.A. Fleury: Phys. Rev. Lett. *37*, 161 (1976)
8.33 R.K. Wehner, R. Klein: Physica (Utrecht) *62*, 161 (1972)
8.34 G. Winterling, W. Senn, M. Grimsditch, R. Katiyar: In *Lattice Dynamics*, ed. by M. Balkanski (Flammarion Sciences, Paris 1978) p.553
8.35 W. Dieterich, P. Fulde, I. Peschel: Adv. Phys. *29*, 527 (1980)
8.36 A.S. Pine: Phys. Rev. *185*, 1187 (1969)
8.37 P. Fulde, H. Wagner: Phys. Rev. Lett. *27*, 1280 (1971)
8.38 J. Pelous: Doctoral Thesis, University of Montpellier, France (1978)
8.39 M. van Haumeder, U. Strom, S. Hunklinger: Phys. Rev. Lett. *44*, 84 (1980)
8.40 O.L. Anderson, H.E. Bömmel: J. Am. Ceram. Soc. *38*, 125 (1955)
8.41 S. Hunklinger: *Proceedings Ultrasonics Symposium* (IEEE, New York 1974) p.493
8.42 D. Ng, R.J. Sladek: Phys. Rev. B*11*, 4017 (1975)
8.43 J.R. Tessman, A.H. Kahn, W. Shockley: Phys. Rev. *92*, 890 (1953)
8.44 I.L. Fabelinskii: *Molecular Scattering of Light* (Plenum Press, New York 1968) Chapter I.4
8.45 P.W. Anderson, B.I. Halperin, C. Varma: Philos. Mag. *25*, 1 (1972)
8.46 W.A. Phillips: J. Low Temp. Phys. *7*, 351 (1972)

Additional References with Titles

H. Böttger: "Vibrational Properties of Noncrystalline Solids", Phys. Status Solidi
  B*62*, 9 (1974)
A.J. Leadbetter: "Vibrational Excitations in Solids", in *International Conference
  on Phonon Scattering in Solids*, ed. by H.J. Albany (La Documentation Francaise,
  Paris 1972) p.338
R.O. Pohl: "Phonon Scattering in Amorphous Solids", in *Phonon Scattering in Solids*,
  ed. by L.J. Challis, V.W. Rampton, A.F.G. Wyatt (Plenum, New York 1976) p.107
R.O. Pohl, G.L. Salinger: "The Anomalous Thermal Properties of Glasses at Low
  Temperatures. Ann. N.Y. Acad. Sci. *279*, 150 (1976)

Subject and Material Index

Applied Physics

A monthly journal

Board of Editors:
S.Amelinckx, Mol; **V.P.Chebotayev,** Novosibirsk;
R.Gomer, Chicago, IL; **P.Hautojärvi,** Espoo;
H.Ibach, Jülich; **K.L.Kompa,** Garching;
V.S.Letokhov, Moskau; **H.K.V.Lotsch,** Heidelberg;
H.J.Queisser, Stuttgart; **F.P.Schäfer,** Göttingen;
R.Ulrich, Hamburg; **W.T.Welford,** London;
H.P.J.Wijn, Eindhoven; **T.Yajima,** Tokyo

Coverage:
Application-oriented experimental and theoretical
physics:

Solid-State Physics	Quantum Electronics
Surface Science	Laser Spectroscopy
Solar Energy Physics	Photophysical Chemistry
Microwave Acoustics	Optical Physics
Electrophysics	Optical Communications

Special Features:
Rapid publication (3-4 months)
No page charges for concise reports
Microform edition available

Languages:
Mostly English

Articles:
Original reports, and short communications
review and/or tutorial papers

Manuscripts:
To Springer-Verlag (Attn. H.Lotsch),
P.O.Box 105 280, D-6900 Heidelberg 1, FRG

Please ask for your sample copy.
Send your order or request to your bookseller
or directly to:
Springer-Verlag, Promotion Department,
P.O.Box 105280, D-6900 Heidelberg 1, FRG
North America: Springer-Verlag New York Inc.,
Journal Sales Dept., 44 Hartz Way, Secaucus,
NJ 07094, USA

Springer-Verlag
Berlin
Heidelberg
New York

Amorphous Semiconductors

Editor: M. H. Brodsky
1979. 181 figures, 5 tables. XVI, 337 pages
(Topics in Applied Physics, Volume 36)
ISBN 3-540-09496-2

Contents:
M. H. Brodsky: Introduction. – *B. Kramer, D. Weaire:*
Theory of Electronic States in Amorphous Semi-
conductors. – *E. A. Davis:* States in the Gap and
Defects in Amorphous Semiconductors. –
G. A. N. Connell: Optical Properties of Amorphous
Semiconductors. – *P. Nagels:* Electronic Transport
in Amorphous Semiconductors. – *R. Fischer:*
Luminescence in Amorphous Semiconductors. –
I. Solomon: Spin Effects in Amorphous Semicon-
ductors. – *G. Lucovsky, T. M. Hayes:* Short-Range
Order in Amorphous Semiconductors. –
P. G. LeComber, W. E. Spear: Doped Amorphous
Semiconductors. – *D. E. Carlson, C. R. Wronski:*
Amorphous Silicon Solar Cells.

O. Madelung

Introduction to Solid-State Theory

Translated from the German by B. C. Taylor
1978. 144 figures. XI, 486 pages
(Springer Series in Solid-State Sciences, Volume 2)
ISBN 3-540-08516-5

Contents:
Fundamentals. – The One-Electron Approxi-
mation. – Elementary Excitations. – Electron-
Phonon Interaction: Transport Phenomena. – Elec-
tron-Electron Interaction by Exchange of Virtual
Phonons: Superconductivity. – Interaction with
Photons: Optics. – Phonon-Phonon Interaction:
Thermal Properties. – Local Description of Solid-
State Properties. – Localized States. – Disorder. –
Appendix: The Occupation Number Represen-
tation.

Glassy Metal I

Ionic Structure, Electronic Transport, and
Crystallization
Editors: H. Beck, H.-J. Güntherodt
1981. 120 figures, 12 tables. Approx. 350 pages
(Topics in Applied Physics, Volume 46)
ISBN 3-540-10440-2

Contents:
H. Beck, H.-J. Güntherodt: Introduction. – *P. Duwez:*
Metallic Glasses-Historical Background. –
T. Egami: Structural Study by Energy Dispersive
X-Ray Diffraction. – *J. Wong:* Exafs Studies of
Metallic Glasses. – *A. P. Malozemoff:* Brillouin Light
Scattering from Metallic Glasses. – *J. Hafner:*
Theory of the Structure, Stability, and Dynamics
of Simple-Metal Glasses. – *P. J. Cote, L. V. Meisel:*
Electrical Transport in Glassy Metals. – *J. L. Black:*
Low-Energy Excitations in Metallic Glasses. –
W. L. Johnson: Superconductivity in Metallic
Glasses. – *U. Herold, U. Köster:* Crystallization of
Metallic Glasses.

The Physics of Selenium and Tellurium

Proceedings of the International Conference on the
Physics of Selenium and Tellurium
Königstein, Fed. Rep. of Germany,
May 28–31, 1979
Editors: E. Gerlach, P. Grosse
1979. 210 figures, 22 tables. X, 281 pages
(Springer Series in Solid-State Sciences, Volume 13)
ISBN 3-540-09692-2

Contents:
Bands and Bonds in Se and Te. – Lattice Dynamics
of Trigonal Se and Te. – Bandstructure in the
Neighbourhood of the Gap of Trigonal Se and Te. –
Imperfections and Impurities in Te. – Transport
Phenomena in Trigonal Se und Te. – The Amor-
phous, Glassy, and Liquid State. – Photoelectric
and Transport Phenomena in Amorphous
Systems. – Crystalline and Amorphous As_2Se_3. –
Preparation and Application. – Index of Contri-
butors.

Springer-Verlag
Berlin
Heidelberg
New York